INTRODUCTION
TO
KINEMATICS

INTRODUCTION TO KINEMATICS

Thomas B. Hardison, P.E.

MECHANICAL ENGINEERING TECHNOLOGY
CATAWBA VALLEY TECHNICAL INSTITUTE

RESTON PUBLISHING COMPANY, INC.
A Prentice-Hall Company
Reston, Virginia 22090

Library of Congress Cataloging in Publication Data

HARDISON, THOMAS B
    Introduction to kinematics.

    Includes index.
    1. Machinery, Kinematics of.   I. Title.
TJ175.H233   621.8'11   79-14070
ISBN 0-8359-3228-1

© 1979 by Reston Publishing Company, Inc.
A Prentice-Hall Company
Reston, Virginia   22090

10   9   8   7   6   5   4   3   2   1

Printed in the United States of America

To Brian and Beth

CONTENTS

PREFACE

A major aim in writing this text is to provide clear and understandable explanations for some of the more difficult concepts involved in the study of kinematics. Experience in teaching the subject matter has shown that the average student has difficulty in understanding the methods used for finding velocities and accelerations, the methods used in analyzing planetary gear trains, and other concepts involved in the study of mechanisms. An effort has been made to provide, by text and illustration, clear explanations for these difficult areas.

Although the text has been written for engineering technology use, it is hoped that it may be of benefit in other areas. Descriptions of mechanisms such as the star wheel and mutilated gears, not normally found in textbooks in kinematics, have been included in addition to most of the common mechanisms. The text does not use calculus but explains the relationship involving calculus in displacement, velocity, and acceleration.

The section devoted to cams includes the conventional layout of cams by graphical means in addition to analytical methods used for simple harmonic motion and cycloidal motion. Chapter 3, on the four-bar linkage, has been introduced early in the text deliberately in order that the student may begin to visualize linkages. Its sequence is not critical, however, and individual instructors may wish to change it.

It is hoped that the aim of providing clarity in the text has been accomplished and that the results will be reflected in the knowledge that the student gains.

ABBREVIATIONS AND SYMBOLS

Symbol or Abbreviation	Meaning
a	Acceleration
a^c	Coriolis acceleration
Alpha (α)	Angular acceleration
a^n	Normal acceleration
a^t	Tangential acceleration
Beta (β)	Angle
C	Number of instant centers
D	Pitch diameter
Delta (Δ)	Change or increment
L	Length
L_c	Center distance
n	Number of ...
N	Revolutions per minute
Omega (ω)	Angular velocity
p	Circular pitch
P	Diametral pitch
Phi (ϕ)	Angle
Psi (ψ)	Helix angle
r	Radius or distance
R	Resultant
s	Displacement
Theta (θ)	Angle
t	Time
v	Velocity
$\leftrightarrow$	Add vectorially
$-\rightarrow$	Subtract vectorially
$>$	Greater than ...

Symbol or Abbreviation	*Meaning*
$<$	Less than . . .
$\geqq$	Equal to or greater than . . .
$\leqq$	Equal to or less than . . .
∞	Infinity

INTRODUCTION
TO
KINEMATICS

CHAPTER 1
AN INTRODUCTION TO KINEMATICS

1-1   HISTORY

Kinematics deals with the motion of machine elements without consideration of any forces acting on the elements. Kinematics may be an unfamiliar term to the reader, but *mechanism* should be a familiar one. Although there is a distinction between the two terms, kinematics and mechanisms are sometimes used interchangeably. A mechanism, however, is the combination of machine elements that provides the motions with which kinematics is concerned.

The history of mechanisms is a long one. The crank and connecting rod mechanism of the automotive internal combustion engine dates from the fifteenth century. In his notes, Leonardo da Vinci showed ideas for the development of mechanisms such as the rack and pinion, link chains, cams, and gears. We know that some of the early clocks date back to the fourteenth century. The development of the escapement, the regulating device used on clocks and watches, is believed to have taken place between 1280 and 1330.

Figures 1-1 and 1-2 illustrate some of these mechanisms. The crank and connecting rod mechanism is shown in Figure 1-1. The great majority of internal combustion engines uses this principle of motion.

Figure 1-2 presents an interesting history of clock mechanisms. It shows the mechanism of a wooden tower clock that was in use about 1830. The mechanism is now in the National Museum of History and Technology in Washington, D.C. With proper workmanship, wooden clocks like this were thoroughly reliable. The poor workmanship of some small clockmakers gave wooden clocks a bad reputation, however. Figure 1-2 shows the escapement and gearing common to watches and clocks.

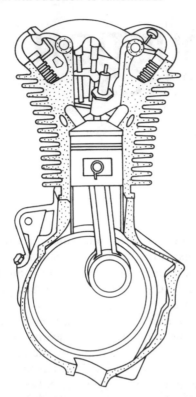

FIGURE 1–1. Sectioned single-cylinder engine, showing crank and connecting rod.

FIGURE 1–2 (below). Wooden clock tower, ca. 1830. The escapement is the small, sharp-toothed wheel near the center and at the top of the mechanism. (Courtesy National Museum of History and Technology, Smithsonian Institution, Photo No. 65181A)

1–2 DEFINITIONS

Before we proceed with the study of kinematics, some important terms must be defined and explained. Kinematics has been partially explained and can now be given a more formal definition.

Kinematics. **"The branch of mechanics that deals with motion in the abstract, without reference to the force or mass."** [1]
Mechanism. **A combination of machine elements or members designed and assembled to produce certain specific motions.**

The *mechanism* can be explained further by making a comparison with the machine, a broader term than mechanism. Machines are usually considered to be made up of mechanisms, but another basis has been made for comparing the two, as follows:

Machine. **A device that does useful work.**
Mechanism. **The combination of machine elements that provides specific motions but that does no useful work.**

The classic example of the mechanism is the watch or clock, whereas the internal combustion engine is an example of the machine. No external work is done by the watch on any object or thing; the internal combustion engine does perform external work.

The mechanism *link* is the mechanical element that is a part of the mechanism. *Link* is a broad term used to describe all components, such as connecting rods or bars, cams, gears, cranks, and belts. All the elements shown in Figure 1–3 are classified as links.

Another term that will be used frequently throughout this text is *inversion*. Mechanisms are normally designed to give certain definite motions to their links. If we constrain, or hold, a link that normally is in motion and at the same time free another link that has been stationary, the mechanism will have the same relative motion between links. From this, we can define inversion.

Inversion. **A change in a mechanism's motion that leaves the relative motion of all links the same, but changes the absolute motion of some of the links.**

Absolute motion means motion with respect to the earth.

---

[1] By permission. From *Webster's New World Dictionary*, Second College Edition. Copyright © 1972 by the World Publishing Co.

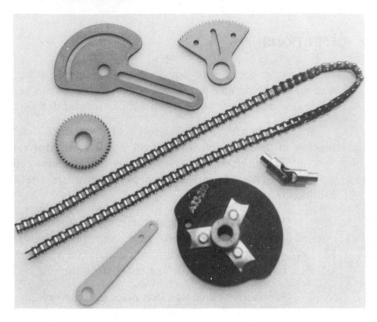

FIGURE 1–3. All these components are links.

1–3 EXAMPLES

There are many examples of common and uncommon machines that use interesting mechanisms. A very common example is the home sewing machine. The World War I aircraft *rotary* engine is a very uncommon machine, which is an inversion of the mechanism in the aircraft *radial* engine. We shall use these three to illustrate the concepts that have been discussed in this chapter.

Figure 1–4 is an exploded isometric drawing of the head-end module of the Singer 900 sewing machine. It shows the complex mechanism used to provide the motions required for the needle, fabric, and thread. There are six other modules, or subassemblies of mechanisms, required for the complete sewing machine.

Figures 1–5 and 1–6 illustrate aircraft *rotary* and *radial* engines, respectively. The Le Rhone rotary engine in Figure 1–5 is mounted in a Standard E-1 advance trainer of World War I. The Continental radial engine in Figure 1–6 is mounted in a replica of the *Spirit of St. Louis* built by the Experimental Aircraft Association Air Museum Foundation. The propeller, crankcase cover, crankcase, and cylinders of the Le Rhone are all assembled and bolted together to form one unit; when the propeller rotates, the complete assembly rotates. In the radial engine the propeller rotates with the

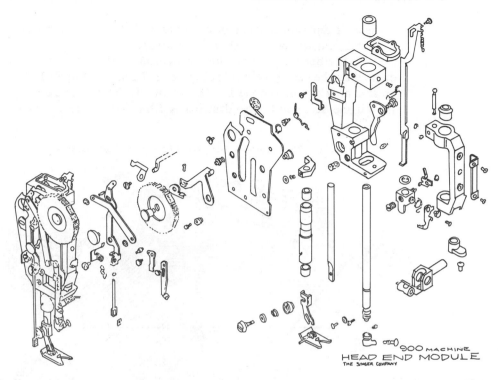

FIGURE 1–4.   (Courtesy The Singer Company)

FIGURES 1–5 and 1–6.   Left: Le Rhone rotary engine of World War I (Courtesy Shannon Air Museum).   Right: Continental radial engine in replica of *Spirit of St. Louis* (Courtesy Experimental Aircraft Association and EAA Air Museum Foundation)

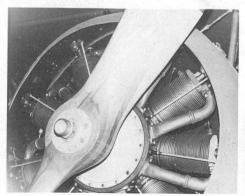

crankshaft, and the other components are stationary. Conversely, in the rotary engine the crankshaft is stationary.

Further insight into the mechanism motion of these two engines can be had by examining Figures 1–7 and 1–8. Figure 1–7 is an assembly drawing of the Le Rhone engine, showing one half of the engine sectioned. The drawing is from a U.S. government

FIGURE 1–7. Assembly drawing of Le Rhone rotary engine.

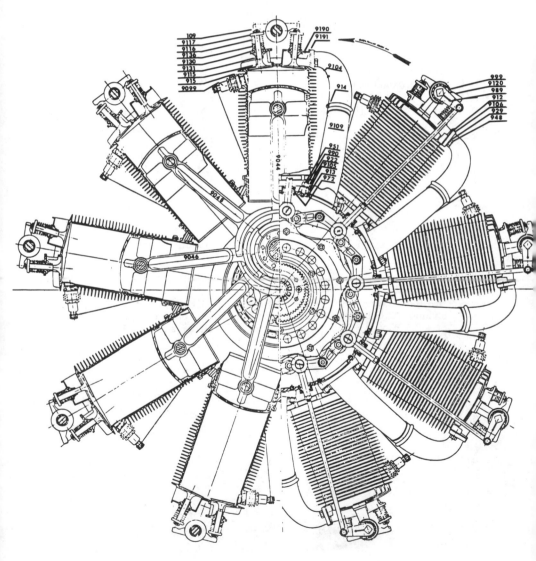

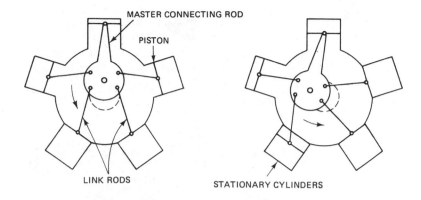

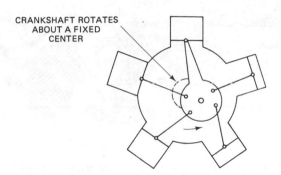

FIGURE 1–8. Functioning of aircraft radial engine.

booklet published in 1919, *Instructions for the 80-Horsepower Le Rhone Engine.*

The functioning of the radial engine is shown schematically by Figure 1–8. A master connecting rod is used and connects to the rotating crankshaft. The link rods all connect to the master rod, as shown. Figure 1–9 shows the crankshaft and connecting rod assembly used in a Pratt and Whitney Wasp engine. The master connecting rod is clearly seen here in the vertical position. The *inversion* of this mechanism would be obtained if the crankshaft were held stationary and the cylinders and crankcase rotated.

1–4 MECHANISM ANALYSIS

The ability to analyze motions in a mechanism gives us, at least to some extent, the capability to design a mechanism for a specific function. This is true in the case of a gear train, for instance.

FIGURE 1–9. Crankshaft and connecting rod assembly of Pratt and Whitney Wasp engine. (Courtesy United Technologies Corporation)

However, in designing a complex linkage in which a specific motion of a point is required, we usually need more than the ability to analyze an existing mechanism. Sometimes trial-and-error solutions are used, and computer techniques have also been developed to aid in the design process. Design (also called *synthesis*) can be a much more difficult process than analysis.

The analysis of mechanism motion not only involves the path taken by points on a particular link; it also is concerned with velocities and acceleration. Velocities are important in machine design because of wear considerations, in addition to many other reasons. Design for adequate strength of machine members is related to acceleration by Newton's second law of motion, which states that

$$\text{force} = \text{mass} \times \text{acceleration}$$

$$F = Ma$$

Although mathematical methods of motion analysis can be used, it will usually be found that graphical methods are easier, faster, and yield sufficient accuracy. Thus, in this text, primary emphasis is on graphical methods, which require a scale layout drawing of the mechanism and depend on vector addition and subtraction.

1–5 SYSTEMS OF UNITS

The physical quantities involved with kinematics are primarily those of length and time, since distance, velocity, and acceleration all have one or more of these as components. In the English system of units, common units for length measurements are the inch and foot; for time, the second or minute is used.

Most industrialized countries have now standardized on the *International System of Units* (SI), or metric system. Use of the SI system in the United States is increasing and its importance is growing. Because of this, both the English and SI systems are used in this text. This involves only the physical quantity of length, since the units for time are the same in both systems.

► *PROBLEMS*

1–1 A common postage scale uses a spring and calibrated lever mechanism to weigh letters. Would you classify the scale as a machine or a mechanism?

1–2 Give an example of an internal combustion engine that does not use the crank and connecting rod mechanism.

1–3 Define kinematics.

1–4 Define mechanism.

1–5 Define inversion and give an example.

1–6 Give two reasons why the determination of velocities in a mechanism is important.

1–7 Mechanisms are commonly thought of as being used in mechanical devices, but they are also used in many electrical devices. Give two examples of electrical devices that depend on a mechanism to function.

1–8 Why is it important to determine accelerations in a mechanism?

1–9 The intake and exhaust valves in the cylinder heads of the Le Rhone rotary engine are operated by the push rods shown on the right side of Figure 1–7. Examine this figure carefully and see if you can determine what operates the push rods.

CHAPTER 2
MECHANISM MOTION

2–1 INTRODUCTION

Motions involved in mechanisms range from the very simple to the extremely complicated. Many simple motions are readily apparent and can be seen by the eye. Some motions may be simple but have velocities so high that analytical techniques have to be used to determine their paths. Other motions are so complicated and have velocities so high that special techniques such as high-speed photography have been used in analyzing them.

In this text we shall be concerned with those motions that can be analyzed with the aid of graphical and mathematical methods. The very complicated motions, particularly those that occur in three dimensions, are beyond the scope of the text.

2–2 PLANE MOTION

A point on a body has *plane motion* when it moves in such a manner that all its positions remain in one plane or in parallel planes. Figure 2–1 illustrates this. Points *A*, *B*, and *C* in Figure 2–1(a) have *plane motion* because the motion of each point is in a plane parallel to the other points. In Figure 2–1(b), points *D* and *E* are in the same plane and have plane motion. Note that points *A*, *B*, and *C* can be projected onto any parallel plane without changing the relative motion of the points.

Figure 2–2 shows a three-dimensional cam. The motion of the follower of this cam is not plane motion. The path of the follower would have components along all three axes shown. In general, this text will deal only with plane motion.

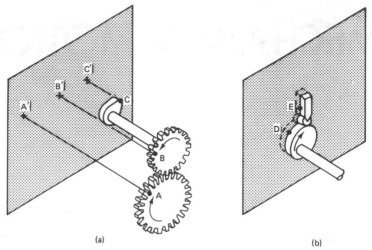

(a)

(b)

FIGURE 2–1. Plane motion.

2–3 TYPES OF PLANE MOTION

There are five types of plane motion: (1) *rectilinear*, (2) *curvilinear*, (3) *translation*, (4) *rotation*, and (5) *combined translation and rotation*. Each type will be defined and explained in order.

Rectilinear motion. **Motion of a *particle* along a straight line. The particle here is assumed to be a body of infinitely small**

FIGURE 2–2. Three-dimensional cam.

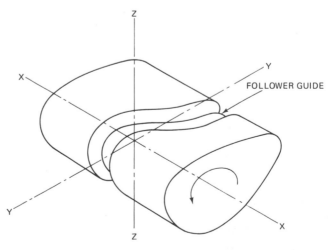

mass; with this in mind, we can think of a point on a larger body as a particle. Figure 2–3(a) illustrates this.

Curvilinear motion. Motion of a *particle* along a curved path. A point on a body can be considered as a particle, as in rectilinear motion. This is shown in Figure 2–3(b).

Note that rectilinear and curvilinear motions are concerned only with *particle* motion. The remaining three motion types are concerned with motions of *bodies.*

Translation. Motion of a *body* such that each *particle* in the body has exactly the same motion. Translation is further subdivided into two types, rectilinear and curvilinear. Examples of rectilinear and curvilinear translation are shown in Figure 2–3(c) and (d).

FIGURE 2–3. Types of plane motion.

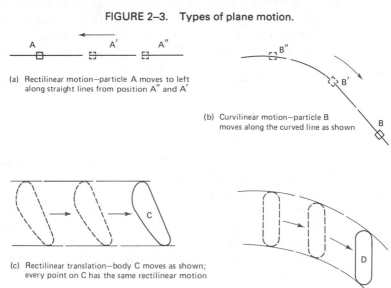

(a) Rectilinear motion—particle A moves to left along straight lines from position A″ and A′

(b) Curvilinear motion—particle B moves along the curved line as shown

(c) Rectilinear translation—body C moves as shown; every point on C has the same rectilinear motion

(d) Curvilinear translation—every point on body D has the same curvilinear motion

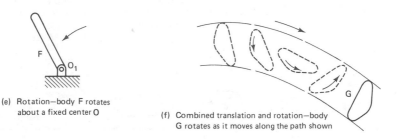

(e) Rotation—body F rotates about a fixed center O

(f) Combined translation and rotation—body G rotates as it moves along the path shown

Rotation. Motion of a body that rotates about a fixed axis, as shown in Figure 2–3(e). Different points on the body have different motions.

Combined translation and rotation. A body moving in such a manner that translation and rotation motion are combined. The translation may be either rectilinear or curvilinear. The example shown in Figure 2–3(f) is curvilinear translation combined with rotation, or rotary motion. Different points on the body again have different motions.

2–4 RELATIVE AND ABSOLUTE MOTION

Relative motion of a point or body is motion of the point or body with respect to some other point or body. *Absolute motion* is defined as motion with respect to the earth. Since the earth travels an orbit, absolute motion as defined here is not truly absolute. However, the definition is adequate for the conventional methods of analyzing motion as used in this text. Both of these types of motion are illustrated in Figure 2–4.

2–5 DISPLACEMENT

Motion implies the changing of the position of an object or point. Three primary characteristics of motion are *displacement, velocity,* and *acceleration.* When these three quantities are determined for a particular motion, we have, in most cases, all the information that we need to define fully the motion in question. A fourth characteristic, *jerk,* has been developed by some authorities in recent years. Jerk is the time rate of change of acceleration. Its use has been limited, and it is beyond the scope of this text.

FIGURE 2–4. Relative and absolute motion. *B* has *relative* motion with respect to *A* and *C* and also has *absolute* motion; *A* and *C* have absolute motion but no motion relative to each other.

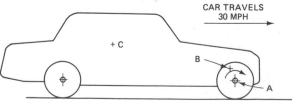

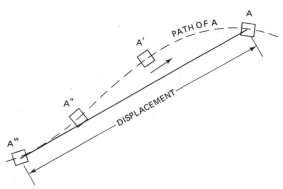

FIGURE 2–5. Displacement of a body.

In this chapter *displacement* is defined and explained. Separate chapters are devoted to *velocity* and *acceleration*. All three characteristics, displacement, velocity, and acceleration, are related, however, and the relationship will be brought out in this chapter. Displacement may be defined as follows:

Displacement.   A change in position of a point or body.

This is a very simple definition but, in order to understand it fully, further explanation is needed. Figure 2–5 shows the path of a body *A* as it moves through successive positions at *A'''*, *A''*, and *A'*. The total distance that *A* travels is represented by the length of the dashed line, which is the path of *A*. The *displacement* of *A* is the straight line drawn between the initial position and the final position. This line then becomes a measure of the change in position of the body *A*. We are not really concerned about the path; any number of different paths could have provided the same displacement as long as the initial and final positions of *A* were the same.

Figure 2–5 shows displacement that would apply to motion of translation as illustrated by Figure 2–3(c) and (c). Rotational motion, in Figure 2–3(e), has *angular displacement* measured by the angle through which the body rotates. In Figure 2–6 a link rotates about a fixed center through the angle. This angle is the angular displacement of the link.

FIGURE 2–6.   Angular displacement.

2–6 VECTOR AND SCALAR QUANTITIES

Scalar quantities are those that can be defined with one number. In a scalar quantity, the number is the magnitude of the quantity and is all that is required to describe the quantity. Seventy degrees describes completely the temperature; thus, temperature is a scalar quantity.

A *vector* quantity requires more than one number for its complete description. Referring to Figure 2–5, the magnitude of the displacement of body *A* can be determined by measuring its length. However, the direction of the displacement has not been specified. To completely define the displacement, both magnitude and direction must be determined. Thus, displacement is a *vector quantity*.

Both of these terms are important in kinematics as well as other fields. A *vector quantity* can be represented by a line with its length proportional to the magnitude and its direction defined by an arrowhead and the angle it makes with some reference plane.

To review, we can write short descriptions of scalars and vectors as follows:

A *scalar quantity* **has magnitude only.**
A *vector quantity* **has magnitude and direction.**

The graphical representation of vectors by using lines allows us to solve problems graphically. Vector representation of different quantities is illustrated in the following example.

EXAMPLE
Draw vectors to represent each of the following quantities:
1. A displacement of 2 feet (ft) due north.
2. A displacement of 1 mile north, 63° east.
3. A displacement of 1.4 meters (m) upward and to the right at an angle of 25° with the horizontal.
4. A displacement of 3.3 centimeters (cm) downward and to the left at an angle of 35° with the horizontal.

Solution
Graphical problems require the use of drawing instruments for accurate solutions. Two triangles, a 30 to 60° and a 45° triangle, are needed, along with a protractor, engineer's scale, and drawing pencils. An additional scale graduated in centimeters is also useful.

The steps in the solution require selecting a suitable scale and then laying off a line of the correct length from some origin. The direction is obtained by measuring the required angle and using an arrowhead to indicate the *sense* of the vector (that is, whether the vector is up or down, or to the right or left). The detailed steps are shown in Figure 2–7, where the horizontal is given as the reference surface for three of the examples, but the fourth uses a magnetic heading. In the case of magnetic headings, the angle is measured from the vertical, or north–south, line.

2–7 VECTOR ADDITION AND SUBTRACTION: GRAPHICAL METHODS

The mathematics of vectors requires that we give consideration to the magnitude and also the direction of the vector. When direction is specified, it implies the use of an angle. Therefore, an angle is

FIGURE 2–7. Vector construction. Steps in drawing a vector:
1. Select a suitable scale—that for (a) is 1 in. = 1 ft.
2. Measure the angle from the reference surface given in the problem
3. Measure the required length of line from the origin and place the arrowhead at the opposite end.

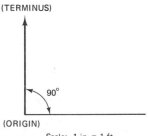

Scale: 1 in. = 1 ft.

(a) Vector representing a displacement of 2 ft. north

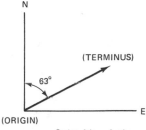

Scale: 1 in. = 1 ml.

(b) Vector representing a displacement of 1 mile north 63° east

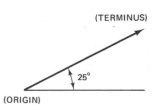

Scale: 1 in. = 1 m.

(c) Vector representing a displacement of 1.4 miles at an angle of 25°

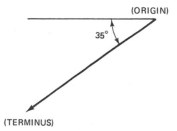

Scale: 1 in. = 1 cm.

(d) Vector representing a displacement of 3.3 cm. at an angle of 35°

always used in some way when doing mathematical operations with vectors.

Vector addition and *subtraction* are the only operations of vector mathematics required in the study of kinematics. Vector addition and subtraction can be done analytically, but in most cases graphical solutions are simpler and provide sufficient accuracy. Both methods are taken up here.

Vector mathematics uses equations to make statements of equality, just as is done in algebra. In this text the symbols used to indicate vector addition and subtraction are different from the plus (+) and minus (−) of conventional algebra. The following symbols are used:

Add vectorially: $+\rightarrow$

Subtract vectorially: $-\rightarrow$

Use of these symbols is not universal, and other textbooks may use different symbols.

Equations are written with these symbols as is done in algebra. Some examples follow:

$$A = B + \rightarrow C$$

$$A = B - \rightarrow C$$

$$Z = X + \rightarrow Y$$

$$Z = X - \rightarrow Y$$

$$A = B + \rightarrow C - \rightarrow D + \rightarrow E$$

Quantities can be transposed from different sides of the equation by changing the sign, as in algebra. Thus, the equation $A - \rightarrow C = B$ can be changed to $A = B + \rightarrow C$. This algebraic method of adding and subtracting vectors is an aid in doing graphical addition and subtraction.

The simplest graphical method of vector *addition* is the *polygon method*. There are other methods, some of which are variations of the polygon method, but the polygon method is the most used. The procedure for the polygon method is as follows:

1. Select one vector and draw it to scale.
2. Select another vector (the order of selection makes no difference) and draw it with its tail (origin) connecting to the head (terminus) of the first.
3. Select the next vector and draw it as in step 2.
4. After the last vector has been drawn in place, draw a line connecting the tail (origin) of the first vector with the

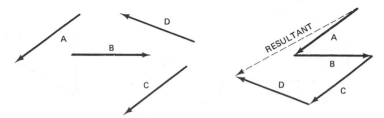

FIGURE 2–8. Graphical addition of vectors: the arrows on all the vectors added run in the same direction in the polygon (in this case, clockwise); the arrow on the *resultant* runs in the opposite direction.

head (terminus) of the last vector. Place an arrowhead on this line next to the arrowhead of the last vector drawn. This line is the *resultant* of the vectors and represents the vector sum of all of them.

This procedure is illustrated in Figure 2–8, and the origin and terminus of a vector are shown in Figure 2–7.

The special case in which the direction of all the vectors is the same can be handled with the polygon method. In this case, we can assume a polygon with all the sides running in the same direction. If all the sides are in the same direction, the vectors to be added are colinear, and the *resultant* is also colinear. Since showing all the vectors and the resultant on the same line would be confusing, we offset the resultant and those vectors of different sense and show them to the side. This is illustrated in Figure 2–9. It is emphasized that the same procedures apply for this case as for the more general case shown in Figure 2–8.

EXAMPLE
Add the vectors shown in Figure 2–10.

Solution
The solution is also shown in Figure 2–10. Starting with vector *A*, the three vectors are laid out head to tail, and the

FIGURE 2–9. Addition of vectors with the same direction.

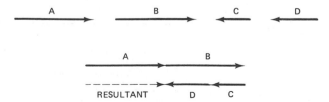

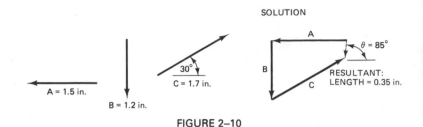

FIGURE 2–10

resistant is drawn connecting the first and last vectors. The order of selection is unimportant; we could have selected the order *C*, *B*, *A*, and the answer would have been the same. Note that the answer is not complete until we have given both the magnitude and the angle it makes with some reference line.

EXAMPLE
Add the vectors shown in Figure 2–11.

Solution
This is the special case in which the direction of the vectors is the same. To make the solution easier, all the vectors to the right (plus) direction are selected first, and then all the vectors to the left (minus) direction are selected. The solution to the problem is shown in Figure 2–11(a) and (b). Part (a) illustrates the simpler procedure of using plus vectors first, and then minus vectors. Part (b) shows the same solution taking the vectors in the order *A*, *B*, and *C*.

FIGURE 2–11

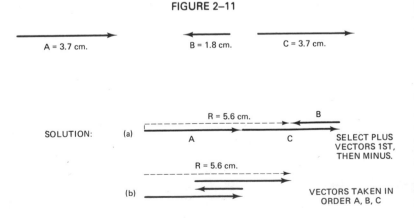

This proves that the order of selection makes no difference in the answer, but in this special case it does provide a simpler method.

Vector subtraction is a little more complicated and requires somewhat more thought to perform correctly. It is a common error to confuse methods for addition and subtraction and to do addition when subtraction is desired. The following points should help you to learn the differences between the two:

1. The *resultant* is always the answer obtained when vectors are *added.*
2. The arrowhead on the *resultant* is always adjacent to the arrowhead of the last vector added.
3. When two vectors are to be *subtracted* graphically, the vectors are placed so that their origins connect, not the head-to-tail connection as used in addition.

Several methods are used to perform vector subtraction. Two methods used for graphical solutions follow.

Method 1

Change the sign of the vector being subtracted and proceed as in vector addition. This method is relatively simple and is probably the easiest to use in most cases. The vector equation previously discussed can be used as an aid in visualizing the steps necessary in the graphical solution. For example, vector B is to be subtracted from C, giving vector A. The equation is

$$C - \rightarrow B = A$$

This can be rewritten as

$$C + \rightarrow (-B) = A$$

The sign of B has been changed in the equation, and B and C are now added to give the resultant A. In the graphical solution, the changed sign of B is obtained by changing the arrowhead to the opposite end of the vector. The vectors are then added graphically by using the normal procedure. This is illustrated in Figure 2–12.

Method 2

Draw the two vectors so that their origins connect. Then draw a vector connecting the heads of the two vectors, with the sense of the

REQUIRED: SUBTRACT B FROM C TO GIVE A

EQUATION:
C − ➤ B = A

FIGURE 2–12. Vector sub-
traction using changed
sign.

PROCEDURE: CHANGE THE SIGN OF B

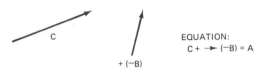

EQUATION:
C + ➤ (−B) = A

ADD:

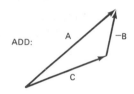

vector from the subtracted vector toward the other vector. This is accomplished by placing the arrowhead next to the arrowhead of the other vector. Using the written equation is also an aid here. Referring to Figure 2–12, the equation is

$$C - \to B = A$$

Now if the $(-B)$ is transposed to the other side of the equation, we have

$$C = A + \to B$$

This is now vector addition, and it will be seen in Figure 2–13 that the vector C now is the resultant of A and B, as the equation says it should be. Referring to Figure 2–12, it will be noted that vector A is the resultant instead of C. This is consistent with the equation used in Method 1, since we said that $A = C + \to (-B)$.

2–8 VECTOR ADDITION AND SUBTRACTION: ANALYTICAL METHODS

There are numerous ways that vector addition and subtraction can be handled analytically. All require the use of trigonometry, and some are more difficult than others. Since most of our work will be

FIGURE 2–13. Vector sub-
traction without chang-
ing sign.

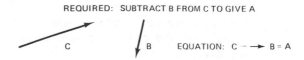

REQUIRED: SUBTRACT B FROM C TO GIVE A

EQUATION: C - ⟶ B = A

PROCEDURE: DRAW THE TWO WITH THE ORIGINS
COINCIDING, THEN CONNECT THE
HEADS OF B AND C. THE SENSE OF
THE VECTOR IS FROM B TO C.

TRANSPOSED EQUATION: C = A + ⟶ B

done graphically, we shall consider here only one analytical method.

All vectors can be resolved into x- and y-components. Once resolved into x- and y-components, all x-components can be added algebraically. The resultant then is described by one x-component and one y-component. These can be vectorially added to provide a single vector.

Consider the vectors shown in Figure 2–14. The x- and y-components of each vector are indicated by the dashed lines in the x- and y-directions. The convention of signs is the same as that in mathematics, with a positive x to the right and a negative x to the left. Similarly, positive y is up and negative y is down. The signs of the components are obtained by using the rules for graphical addition; that is, the vector is the resultant of the two components, and the arrowheads are placed on the components in accordance with the rules previously described.

FIGURE 2–14

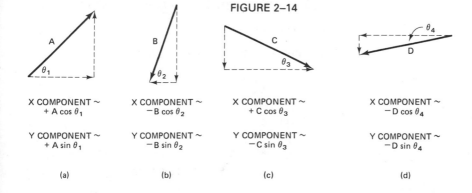

X COMPONENT ~
+ A cos θ_1

Y COMPONENT ~
+ A sin θ_1

(a)

X COMPONENT ~
− B cos θ_2

Y COMPONENT ~
− B sin θ_2

(b)

X COMPONENT ~
+ C cos θ_3

Y COMPONENT ~
− C sin θ_3

(c)

X COMPONENT ~
− D cos θ_4

Y COMPONENT ~
− D sin θ_4

(d)

Each geometric figure in Figure 2–14 is a right triangle. Therefore, we can write algebraic terms for the components in terms of the vector magnitude and an angle. These algebraic terms are shown for the vectors in Figure 2–14. The angle has been selected to be measured from the horizontal, but the other angle (from the vertical) could have been used as well.

We can now add algebraically the components, taking each direction one at a time. The equations are written in this manner:

$$x = A \cos \theta_1 - B \cos \theta_2 + C \cos \theta_3 - D \cos \theta_4 \qquad (2–1)$$

$$y = A \sin \theta_1 - B \sin \theta_2 - C \sin \theta_3 - D \sin \theta_4 \qquad (2–2)$$

When solved, x and y become the sides of a right triangle, and the hypotenuse is the resultant vector.

EXAMPLE

Given the following values for the vectors in Figure 2–14, find the resultant of all the vectors analytically.

$$A = 27 \text{ cm}, \quad \theta_1 = 48°$$

$$B = 24 \text{ cm}, \quad \theta_2 = 72°$$

$$C = 30 \text{ cm}, \quad \theta_3 = 25°$$

$$D = 28 \text{ cm}, \quad \theta_4 = 11°$$

Solution

Equations 2–1 and 2–2 are used in the solution. Substituting, the x-component is

$$x = 27 \cos 48° - 24 \cos 72° + 30 \cos 25° - 28 \cos 11°$$

$$= 10.35 \text{ cm}$$

The y-component is

$$y = 27 \sin 48° - 24 \sin 72° - 30 \sin 25° - 28 \sin 11°$$

$$= -20.78 \text{ cm}$$

We now have the two components in Figure 2–15. The

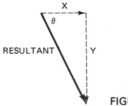

FIGURE 2–15

resultant is also shown. Since the resultant is the hypotenuse of a right triangle, and in a right triangle the hypotenuse equals the square root of the sum of the squares of the two sides, then

$$\text{resultant } R = \sqrt{x^2 + y^2}$$
$$= \sqrt{(10.35)^2 + (20.78)^2}$$
$$= 23.2 \text{ cm}$$

The angle θ is obtained as follows:

$$\theta = \tan^{-1}\left(\frac{y}{x}\right)$$
$$= \tan^{-1}\left(\frac{-20.78}{10.35}\right)$$
$$= -63.5°$$

The sense of the resultant is shown by the arrowhead drawn on it in Figure 2–15. If a figure is not drawn to represent the resultant, a statement should be made describing the sense, such as "down and to the right." It is important to note in this solution that Equations 2–1 and 2–2 were written to describe only the vectors in Figure 2–14 and that they are not general equations. Each set of vectors requires its own equations and component terms.

2–9 KINEMATIC SYMBOLS

Mechanisms are represented in schematic form in somewhat the same way as are electronic circuits. This requires the use of special symbols, each of which has a specific meaning. In general, the symbols show that a certain type of motion exists or indicate that certain links of a mechanism are stationary. The commonly used symbols are shown, with explanation, in Figure 2–16.

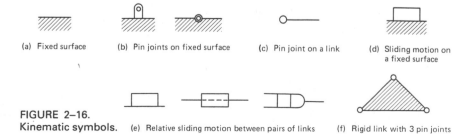

FIGURE 2–16.
Kinematic symbols.

(a) Fixed surface

(b) Pin joints on fixed surface

(c) Pin joint on a link

(d) Sliding motion on a fixed surface

(e) Relative sliding motion between pairs of links

(f) Rigid link with 3 pin joints

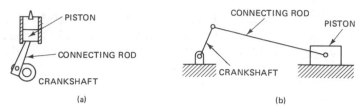

(a)   (b)

FIGURE 2–17. Schematic representation of a mechanism.

How these symbols are used can be seen by examining Figure 2–17, which shows a sectional view of the crankshaft, connecting rod, piston, and cylinder of a one-cylinder internal combustion engine. Beside it is the schematic that represents the same mechanism. The schematic has been rotated 90°, which has no special significance other than that, conventionally, this type of mechanism is shown horizontally.

2–10 DISPLACEMENT, VELOCITY, AND ACCELERATION RELATIONSHIP

Displacement has previously been defined as a change in position of a body or point. When *time* is used to measure how long it takes for a certain displacement to occur, we have *velocity*. When time is used in measuring velocity changes, the result is *acceleration*. The definitions of velocity and acceleration are as follows:

Velocity. The rate of change of *displacement* with respect to time.
Acceleration. The rate of change of *velocity* with respect to time.

A simplified mathematical explanation is necessary to understand these definitions. In Figure 2–18, a point P moves along a curve to successive positions at P_1, P_2, P_3, and P_4. In each position a vector is drawn representing each *increment*, or change, in displacement. Each increment represents the displacement from the

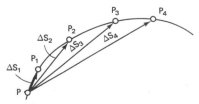

FIGURE 2–18

initial position at P to each of the four positions, P_1, P_2, P_3, and P_4. The displacement Δs_4 is the displacement from P to P_4, for example. We use the symbol Δ (delta) to indicate the increment, or change, in quantity.

For each increment of displacement, there is a corresponding increment of time, Δt, in which the displacement occurs. If we let the displacement Δs_4 equal 12 inches (in.), and the time interval Δt_4 equal 3 seconds (sec), then

$$\frac{\Delta s_4}{\Delta t_4} = \frac{12 \text{ in.}}{3 \text{ sec}} = 4 \text{ in./sec, the rate of change of}$$

displacement with respect to time

Letting

$$v = \text{velocity}$$

$$v_4 = \text{velocity of the point during displacement } s_4$$

then

$$v_4 = \frac{\Delta s_4}{\Delta t_4} = 4 \text{ in./sec}$$

The velocity v_4 is an *average velocity* obtained for the relatively large displacement Δs_4. If we progressively reduce the size of the displacements, as shown by Δs_3, Δs_2, and Δs_1, it can be shown mathematically that the term $\Delta s/\Delta t$ approaches the *instantaneous* velocity as Δt approaches zero. A rigid mathematical analysis requires the use of calculus to a greater degree than desirable for this text; therefore, we shall only show a simplified relationship of velocity and acceleration to calculus.

In calculus the rate of change of one variable with respect to another is given by the *derivative* of one variable with respect to the other variable. If we use s and t for the variables, the derivative is written ds/dt. It is obtained by letting increments approach zero, just as we did with the increment Δt. Since we previously defined velocity and acceleration as the rate of change of two variables, we can make these important statements:

**Velocity is the derivative of displacement with respect to time.
Acceleration is the derivative of velocity with respect to time.**

Mathematically, these statements are written

$$v = \frac{ds}{dt} \tag{2-3}$$

$$a = \frac{dv}{dt} \tag{2-4}$$

Using these equations, the v and a terms represent instantaneous values of velocity and acceleration.

Although none of the problems in this text requires the use of calculus, it is important to remember the basic relationships that exist.

▶ PROBLEMS

2–1 Figure 2–19 shows a cam with a roller follower. The cam rotates in the direction shown, and its axis of rotation is fixed.

    (a) What type of motion does the cam have?

    (b) What type of motion does the roller have?

    (c) What type of motion does the follower have?

    (d) At the point of contact, is there relative motion between a point on the roller and a point on the cam? Give an explanation for your answer.

    (e) Is there relative motion between the roller and the follower?

2–2 Give an example of a body that has combined motion of translation and rotation.

2–3 Identity the scalar quantities and the vector quantities in the following physical quantities.

    (a) force   (e) velocity

    (b) displacement   (f) acceleration

    (c) temperature   (g) time

    (d) energy   (h) mass

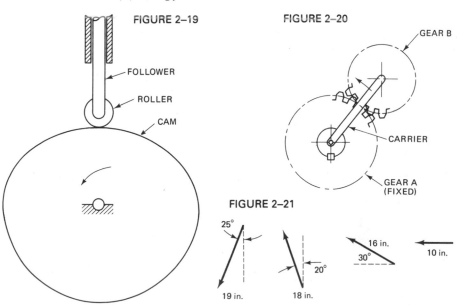

FIGURE 2–19

FIGURE 2–20

FIGURE 2–21

FIGURE 2–22

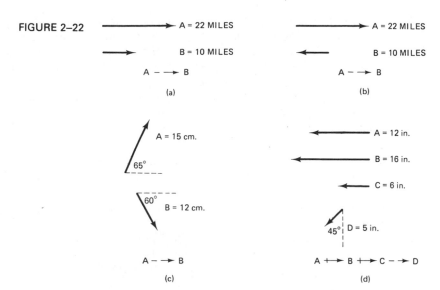

(a)

(b)

(c)

(d)

2–4 The mechanism in Figure 2–20 is called a planetary gear train. Gear *A* is fixed in space and the carrier turns counterclockwise (ccw). What type of motion does gear *B* have?

2–5 Add the vectors in Figure 2–21 graphically.

2–6 Repeat Problem 2–5 by doing the operation analytically, using *x*- and *y*-components. Find the resultant and compare your answer with the answer obtained graphically.

2–7 Perform the indicated operations in Figure 2–22, first using graphical methods and then analytical.

2–8 In Figure 2–23, vector *R* is the resultant of vectors *A* and *B*. Find vector *B*.

2–9 A man in a rowboat starts to row across a stream. The river runs directly north and south at the point where he starts. He starts rowing directly west, but finds that he is $\frac{1}{2}$ mile south of his starting point when he reaches the other bank. If the river is $1\frac{1}{4}$ miles wide, what is his displacement?

2–10 Find the north and west components of a displacement of 63 miles north 63° west. Solve graphically and analytically.

2–11 A point *P* moves as shown in Figure 2–24 from position P_0 to position P_2. What is its final displacement?

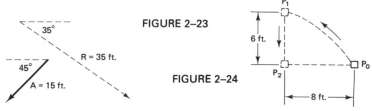

FIGURE 2–23

FIGURE 2–24

2–12 In Figure 2–25, *R* is the resultant of vectors *A, B, C,* and *D*. Vector *C* makes an angle of 85° with the horizontal, and vector *D* is horizontal. Find *C* and *D* graphically. (*Hint:* The point where *C* and *D* meet is the intersection of lines in the directions given for *C* and *D*.)

2–13 At top of stroke the instantaneous velocity of the piston in an internal combustion engine is zero. If the time required for the piston to travel from the top of its stroke to the bottom of the stroke is 0.02 sec, and the stroke is 4 in., what is the average velocity of the piston?

2–14 The mechanism shown in Figure 2–26 is called a *quick-return* mechanism. Link *F* is the fixed member. Identify the pairs of links that have relative sliding motion.

2–15 Add the vectors shown in Figure 2–27 graphically. Check your answer by doing the problem analytically.

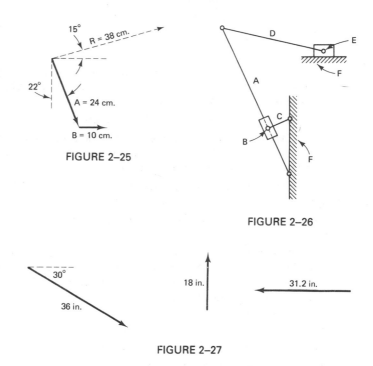

15°
R = 38 cm.
22°
A = 24 cm.
B = 10 cm.

FIGURE 2–25

D
E
A
F
C
B
F

FIGURE 2–26

30°
36 in.
18 in.
31.2 in.

FIGURE 2–27

CHAPTER 3
THE FOUR-BAR LINKAGE

3-1 INTRODUCTION

An important class of mechanisms is the *four-bar linkage.* It is essentially an assembly of four links, or bars, connected together by pin joints so that relative rotational motion can exist between adjoining links. The uses of four-bar linkages are so numerous that we could not list them all. Consider the old steam locomotive sketched in Figure 3–1. The driving wheels are linked together by a coupler so that the tractive effort of the locomotive is shared by both pairs of wheels. The two wheels, coupler, and the frame of the locomotive constitute a four-bar linkage, which can be represented by the schematic of Figure 3–2. The driver mechanism (the piston, connecting rod, and front wheel) is an example of a slider crank mechanism, which we shall consider in Chapter 7.

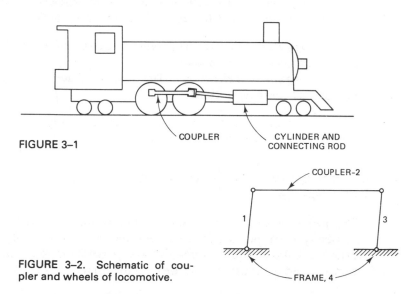

FIGURE 3–1   COUPLER   CYLINDER AND CONNECTING ROD

FIGURE 3–2. Schematic of coupler and wheels of locomotive.

The four-bar mechanism in Figure 3–2 consists of links 1 and 3, of equal length, the coupler 2, and the frame 4. Note that there are two pin joints on the frame, but we consider the frame as only one link. This is true of all mechanisms; no matter how many points of the frame are connected to other links, the frame is only one link of the mechanism. The frame in this case is the body or chassis of the locomotive.

3–2 ANGULAR DISPLACEMENT

The wheels of the locomotive in Figure 3–1 rotate with respect to the locomotive. In Figure 3–2, links 1 and 3 rotate with respect to the frame. As noted briefly in Chapter 2, the displacement of an object having rotary motion is described as *angular* displacement. Its measurement is in degrees, radians, or revolutions. In using radians, it is helpful to review the following definition:

Radian. **The central angle subtended by a circular arc whose length is equal to the radius of the circle.**

Since the circumference of a circle is equal to $2\pi \times$ the radius, it follows that there are 2π radians (rad) in 360°, or 1 revolution (rev). Stated mathematically,

$$2\pi \text{ rad} = 360° = 1 \text{ rev} \qquad (3\text{–}1)$$

Angular velocity and angular acceleration have the same general relationships to angular displacement as those given for linear displacements in Chapter 2. We define them as follows:

Angular velocity. **The rate of change of *angular* displacement with respect to time.**
Angular acceleration. **The rate of change of *angular* velocity with respect to time.**

Angular displacements will be used frequently in analyzing the motions involved in four-bar linkages.

3–3 MOTION VARIATION IN FOUR-BAR MECHANISMS

The almost infinite possibilities of different motions available in certain types of mechanisms have been mentioned previously. The four-bar linkage is an example of the possibilities that may be encountered. To illustrate this, we refer to Figure 3–2. To obtain the

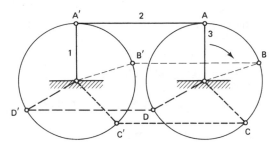

FIGURE 3–3

motion needed to drive the wheels of the locomotive, we know almost instinctively that the lengths of links 1 and 3 have to be the same. This could be proved mathematically, but can be seen more easily in the layout of Figure 3–3. In this layout the links are shown in four different positions as they rotate about the fixed axes of 1 and 3. The coupler 2 remains horizontal in all positions; thus, it has translational motion but no rotational motion.

Now let us consider what happens if we make the lengths of 1 and 3 unequal, as is illustrated in Figure 3–4. Link 1 is shown in three different positions. Links 2 and 3 are shown also in the positions corresponding to the three positions of link 1. In positions *A* and *B* it is noted that link 3 is not parallel as in Figure 3–3. Also, note that in position *C* link 3 cannot be connected to 2 because of insufficient length. This, then, is an impossible condition that could not occur in practice. Since link 2 does not remain parallel, it has combined rotational and translational motion.

The example illustrated by Figure 3–4 is shown to demonstrate the many combinations of linkages that may exist simply by changing dimensions of one or more links in the mechansim. Although we cannot define or analyze all the motions that can occur with all possibilities of four-bar linkages, it is possible to set up parameters governing certain *classes* of four-bar linkages. These

FIGURE 3–4

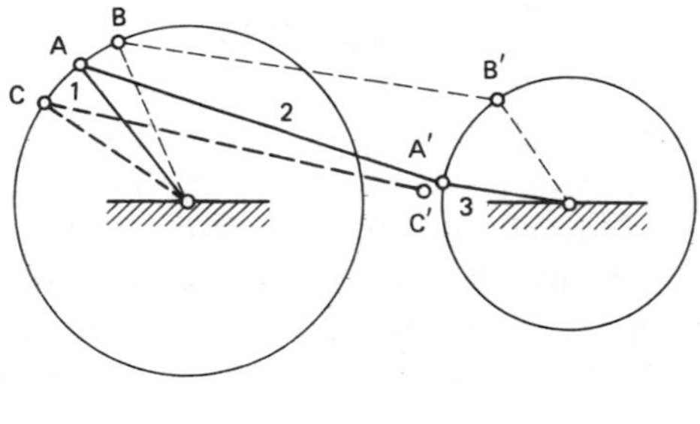

parameters involve the relationships among the lengths of the various links.

Consider first the link lengths that are necessary to form a four-bar linkage. Figure 3–5(a) shows a four-bar linkage in which link 1 is identified as the crank, or driver, link 2 is the coupler, and link 3 is the lever. Link 3 acts as a follower to the input motion of link 1.

Now let us change the length of link 4, the frame, by increasing it to the length shown in Figure 3–5(b). If links 1, 2, and 3 are stretched out in a straight line, we see that they do not connect. Thus, it is impossible to make a four-bar linkage with the lengths shown. From this example, we can make the following statement.

A four-bar linkage cannot be made if one of the links has a length greater than the sum of the other three.

Using Figure 3–5(b), we can express this mathematically, using L to represent length.

$$L_1 + L_2 + L_3 > L_4 \qquad (3-2)$$

The symbol $>$ means "greater than." In the discussions that follow, this symbol and similar ones are used frequently. These symbols and their meanings are shown in the following list.

Symbol	Meaning
$>$	Greater than
$<$	Less than
$\geq$	Equal to or greater than
$\leq$	Equal to or less than

Note that the length of link 4 is the distance between the two pin joints on the frame.

3–4 CRANK ROCKER MECHANISM

The first class of four-bar linkages that we shall consider is called the *crank rocker mechanism*. In this mechanism the crank is capable of rotating through a full 360°, while the lever, as shown in Figure 3–5, can only rotate through a restricted length of arc. The lever then produces oscillatory motion in this arc, and, for this mechanism, is called the *rocker*.

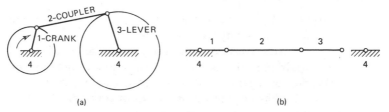

FIGURE 3–5

Figure 3–6 shows a crank rocker mechanism with the links in two different positions. Using the scale graphical constructions in Figure 3–6(a) and (b), we shall develop the parameters, or conditions, that determine a crank rocker mechanism.

In examining Figure 3–6(a), it is apparent that link 1 is the shortest link. Link 1 is also the crank, and a general statement can be made that, in all crank rocker mechanisms, the shortest link is the crank. Although this can be proved mathematically, we shall omit the proof here and accept the statement as fact.

The other conditions involve the lengths of the other links. In Figure 3–6(a), the linkage is in a *limiting* position. Point A can move no farther to the right and must reverse its direction as link 1 rotates. For this position, we can make the following mathematical statement.

$$L_1 + L_2 < L_3 + L_4 \qquad (3\text{–}3)$$

Inequality 3–3 says that the combined lengths of links 1 and 2 are less than the combined lengths of links 3 and 4. This, of course, has to be true since 1 and 2 form a straight line connecting two points. Inequalities can be handled with some of the algebraic rules pertaining to equations. Quantities may be moved from one side to the other by changing the sign. Performing an operation like this gives the following inequality:

$$L_1 + |L_2 - L_3| < L_4 \qquad (3\text{–}4)$$

FIGURE 3–6.   Crank rocker mechanism.

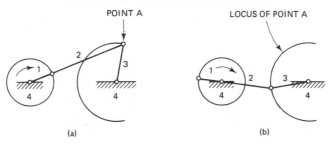

The term $|L_2-L_3|$ means the absolute value of the difference, or the number without any sign prefixed.

Now consider the position of the mechanism in Figure 3–6(b). In this position, link 1 coincides with the position of link 2, and we can again make some observations about the lengths of links. The distance from the center of rotation of link 1 to link 3 is equal to L_2-L_1. With this we can make the following inequality:

$$L_4 < L_2 - L_1 + L_3 \qquad (3\text{--}5)$$

Inequalities 3–4 and 3–5 can be combined into one statement, as follows:

$$L_1 + |L_2 - L_3| < L_4 < L_2 - L_1 + L_3 \qquad (3\text{--}6)$$

The requirements of the *crank rocker mechanism* can now be summarized:

1. **The shortest link is always the crank.**
2. **The lengths of the links must conform to Inequality 3–6, repeated here.**

$$\boldsymbol{L_1 + |L_2 - L_3| < L_4 < L_2 - L_1 + L_3} \qquad \textbf{(3--6)}$$

3–5 DRAG LINK MECHANISM

When both the crank and the lever of a four-bar linkage are capable of rotating through 360°, the linkage is called a *drag link mechanism*. A *drag link mechanism* is shown in Figure 3–7. The shortest link is link 4, the fixed link or frame. This condition is necessary for all drag link mechanisms; so we can say that one of the requirements of the drag link mechanism is that the shortest link be always the fixed link.

Now consider the other positions of the mechanism, as shown in Figure 3–8. First using Figure 3–8(a), we can set up the

FIGURE 3–7. Drag-link mechanism.   FIGURE 3–8

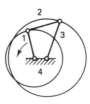

(a)   (b)

following statement:

$$L_3 + L_4 < L_1 + L_2 \qquad (3\text{--}7)$$

Rearranging,

$$L_3 < L_1 + L_2 - L_4 \qquad (3\text{--}8)$$

In Figure 3–8(b), link 3 is rotated 180° from the position in Figure 3–8(a). This position gives another requirement in the lengths of the links, as follows:

$$L_2 < L_1 + L_3 - L_4 \qquad (3\text{--}9)$$

Rearranging,

$$|L_2 - L_1| + L_4 < L_3 \qquad (3\text{--}10)$$

The absolute value of $L_2 - L_1$ is used, since we do not know which link is larger. Inequalities 3–8 and 3–10 can be written as one statement also.

$$|L_2 - L_1| + L_4 < L_3 < L_1 + L_2 - L_4 \qquad (3\text{--}11)$$

The requirements of the *drag link mechanism* are then summarized as follows:

1. **The shortest link is always the fixed link, or frame.**
2. **The lengths must conform to Inequality 3–11, repeated here.**

$$|L_2 - L_1| + L_4 < L_3 < L_1 + L_2 - L_4 \qquad (3\text{--}11)$$

An important characteristic of the drag link mechanism is the varying angular velocity of the follower, which exists when the crank rotates at a constant angular velocity.

3–6 DOUBLE ROCKER MECHANISM

In the *double rocker mechanism* neither the crank nor follower is capable of making a full revolution. Thus, both the crank and follower oscillate back and forth between finite angular limits. A double rocker mechanism is shown in Figure 3–9, where the paths taken by points A and B are indicated by the arcs drawn through each point. Note that each arc has a limited length determined by locating the extreme, or limiting, positions of points A and B. An example of how this is done is shown in Figure 3–10. Here point A has moved as far to the left as it can; link 2 and 3 form one straight line in this position and prevent A from moving any farther. The extreme right position of A is determined similarly.

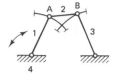

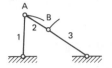

FIGURE 3–9. Double-
rocker mechanism.

FIGURE 3–10. Limiting
position of a double-
rocker mechanism.

In Figure 3–9, it is apparent that the shortest link is link 2. This is the necessary condition to establish a double rocker mechanism, and we can now make the following statement:

For a *double rocker mechanism* to exist, the coupler must be the shortest link.

If the requirement that the coupler be the shortest link is met, there is no need for the statements of inequality required for the other classes of four-bar linkages.

3–7 INDEFINITE MOTION IN FOUR–BAR LINKAGES

If the inequality signs of Inequalities 3–6 and 3–11 are replaced with equality signs (=), the motions of the mechanisms may become indefinite. Although we do not normally consider the effect of forces acting on the mechanism, forces caused by inertia or by external loading can cause a four-bar linkage to behave in a different manner than normal. If the crank in Figure 3–5(a) is a heavy flywheel, there are additional inertial loads on the crank rocker mechanism. Conditions like these may then combine to make the linkage motion indefinite.

The *precision* of the mechanism design is also a factor in proper functioning of the linkage. Loose bearings, excessive play in pin joints, and high frictional forces may all cause unintended motions in the linkage.

3–8 DESIGN OF FOUR-BAR LINKAGES

The example shown here illustrates how the parameters that have been discussed in this chapter are used to design four-bar linkages.

EXAMPLE
A crank rocker mechanism is required for the mechanism of an oscillating sprinkler. The crank is driven by a hydraulic motor that rotates under water pressure. It has been

determined that the available torque from the motor dictates the use of a crank arm $\frac{5}{8}$ in. long. Determine suitable lengths for the other links of the mechanism.

Solution

Figure 3–11(a) is a schematic representation of the mechanism, and Figure 3–11(b) is a sketch of the mechanism. Link 1 is the crank, and link 4 represents the body of the sprinkler. Designs of this type of sprinkler usually use perforated tubing at A to obtain the spray; in Figure 3–11, the tubing at A would be perpendicular to the plane of the paper. As A oscillates, the spray of water moves over an area of ground.

In a crank rocker mechanism, the shortest link is the crank. Inequality 3–6 must also be met. Repeated here, Inequality 3–6 is

$$L_1 + |L_2 - L_3| < L_4 < L_2 - L_1 + L_3 \qquad (3\text{–}6)$$

The solution is essentially obtained by a trial and error. To simplify it, we shall set up the following tabular form for the data.

Assumed Values in Inches

	Trial 1	Trial 2	Trial 3
L_1	$\frac{5}{8}$	$\frac{5}{8}$	$\frac{5}{8}$
L_2	1	$1\frac{1}{4}$	$1\frac{7}{8}$
L_3	$1\frac{1}{4}$	$1\frac{1}{4}$	$1\frac{7}{8}$
L_4	2	2	$1\frac{3}{4}$
$|L_2 - L_3|$	$\frac{1}{4}$	0	0
$L_2 - L_1 + L_3$	$1\frac{5}{8}$	$1\frac{7}{8}$	$3\frac{1}{8}$

FIGURE 3–11. Lawn-sprinkler mechanism.

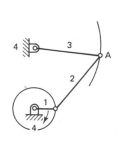

(a) Schematic

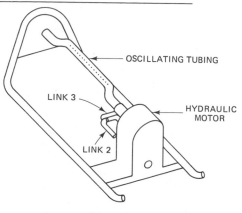

OSCILLATING TUBING

LINK 3

HYDRAULIC MOTOR

LINK 2

(b) Sketch of sprayer

Inequality 3–6 says that L_4 must be less than the term $L_2 - L_1 + L_3$. In the first and second trials this is not true; however, the term $|L_2 - L_3|$ is satisfactory, since it is less than L_4 in all cases. By increasing the values of L_2 and L_3 and decreasing L_4, we arrive at a satisfactory solution in trial 3.

This is not the only solution that will satisfy the requirements. In practice, most designs provide an adjustment to change the spray angle. This is accomplished by providing adjustments that change the length of one link.

3–9 COUPLER CURVES

The motion of a point on the coupler of a four-bar mechanism is of interest. The path traced by a point on the coupler of a four-bar mechanism is sometimes used to perform specific functions. Figure 3–12 shows the plot of a point A that has been selected at the midpoint of the coupler of the double rocker mechanism. The plot is made by drawing the coupler in different positions as the mechanism assumes different positions. To avoid confusion, all the construction lines have been omitted from Figure 3–12, and only the path of point A has been shown.

Since an infinite number of coupler curves may be obtained, design for a specific curve is somewhat difficult. A design aid is an atlas of coupler curves, *Analysis of the Four-Bar Linkage*, by John A. Hrones and George L. Nelson.[1] Digital computers have

[1] Hrones, John A., and George L. Nelson, *Analysis of the Four-Bar Linkage*, Cambridge, Mass.: M.I.T. Press; and New York, N.Y.: John Wiley & Sons, Inc., 1951.

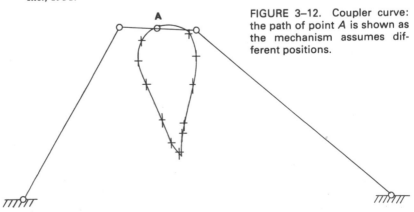

FIGURE 3–12. Coupler curve: the path of point A is shown as the mechanism assumes different positions.

also been used to aid mechanism design. One such method is described by R. W. Mann in *Computer-Aided Mechanism Design*, American Society of Mechanical Engineers paper No. 64-Mech-36, 1964.

3–10 SUMMARY

In this chapter we have introduced the first analytical examination of an important group of mechanisms. The applications of the four-bar linkage to other mechanisms with more specific uses will be taken up in later chapters, where we shall also discuss velocities and accelerations of points and links of four-bar linkages.

► PROBLEMS

3–1 In the locomotive referred to in Figure 3–1, what is the relationship of the distance from the center of one driving wheel to the point where the coupler is attached and the corresponding distance on the second driving wheel?

3–2 Analyze the lengths of links given in the following examples, and determine if it is possible to make four-bar linkages from them.

        (a) $L_1 = 8$ in., $L_2 = 3$ in., $L_3 = 2$ in., $L_4 = 1$ in.
        (b) $L_1 = 2$ in., $L_2 = 3$ in., $L_3 = 5$ in., $L_4 = 8$ in.
        (c) $L_1 = 3$ in., $L_2 = 7$ in., $L_3 = 2$ in., $L_4 = 1$ in.
        (d) $L_1 = 7$ in., $L_2 = 3$ in., $L_3 = 2$ in., $L_4 = 2$ in.

3–3 A crank rocker mechanism has a crank 6 in. long. The length of the fixed link is 9.5 in. Determine suitable lengths for the other two links.

3–4 Make a scale drawing schematic of the crank rocker mechanism determined in Problem 3–3. By plotting different positions of the crank, determine the extreme positions of the follower. Measure the angle the follower makes as it oscillates back and forth from the extreme positions.

3–5 The crank rocker mechanism, in some cases, depends on inertial forces to drive the crank through the extreme position (sometimes called a *dead position*), as shown in Figure 3–13. Name one way

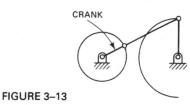

FIGURE 3–13

that this can be provided in the design of the crank (design in this case means the complete part design, including length of the link, material selection, size, and shape).

3–6 In a proposed crank rocker mechanism, the crank length has been selected as 7.5 in., the coupler length is 9 in., and the follower length is 15 in. The crank and follower are mounted on the frame in bearings 8 in. apart. Determine the design feasibility.

3–7 In a drag link mechanism the frame length is 6 in. and the coupler length is 10 in. Determine suitable lengths for the other two links.

3–8 The design of a drag link mechanism shows that the fixed link has a length of $6\frac{1}{4}$ in., and the coupler has a length of 7 in. Lengths of the other two links are $7\frac{1}{4}$ and $8\frac{1}{4}$ in. Check if the lengths are correct for a drag link mechanism.

3–9 A double rocker mechanism has the following lengths for the links:

$$L_1 = 3.2 \text{ in.,} \qquad L_2 \text{ (coupler)} = 1.25 \text{ in.}$$
$$L_3 = 4.2 \text{ in.,} \qquad L_4 \text{ (fixed link)} = 6 \text{ in.}$$

Determine the number of degrees the coupler rotates in one complete cycle of operation. (*Hint*: Find graphically by drawing the linkage to scale, locating various positions of the links, and noting how the coupler moves.)

3–10 Determine the type of four-bar linkage for each of the following mechanisms. In all cases L_3 is the follower and L_4 is the fixed link.

(a) $L_1 = 2\frac{1}{2}$ in., $L_2 = 1$ in., $L_3 = 2\frac{1}{2}$ in., $L_4 = 3$ in.
(b) $L_1 = 15$ mm., $L_2 = 15$ mm, $L_3 = 10$ mm, $L_4 = 7$ mm.
(c) $L_1 = 8$ cm., $L_2 = 10$ cm., $L_3 = 10$ cm, $L_4 = 12$ cm.
(d) $L_1 = 9$ in., $L_2 = 11$ in., $L_3 = 12$ in., $L_4 = 11$ in.

CHAPTER 4
VELOCITY

4–1 INTRODUCTION

In Chapter 2 it was shown that three important characteristics of motion are displacement, velocity, and acceleration. Velocity was defined in Chapter 2, and its definition is repeated here.

Velocity. **The rate of change of displacement with respect to time.**

Velocity is a vector quantity and therefore must be handled mathematically by vectors.

The methods for determining velocities of points and bodies discussed here are used for all types of mechanisms.

4–2 LINEAR VELOCITY

Linear velocity relates to the time rate of change of linear displacement of a point or body. In normal usage when the term *velocity* is used, linear velocity is intended. *Angular velocity* relates to the time rate of change of angular displacement and is always referred to as angular velocity. When a body has angular velocity, the velocities of points on the body can be expressed in linear velocity terms. When there is no angular velocity, the only velocity that exists is linear.

In Chapter 2 an equation for the *average velocity* was developed for a specific set of conditions (shown in Figure 2–18). This equation holds for all conditions and is repeated in a generalized form here.

$$\bar{v} = \frac{\Delta s}{\Delta t}$$

(4–1)

where $\qquad$ $\bar{v} =$ average velocity

$\Delta s =$ increment of displacement

$\Delta t =$ increment of time

Realizing that the s and t terms in Equation 4-1 are always increments, we can drop the delta (Δ) prefix and rewrite the equation as follows:

$$\bar{v} = \frac{s}{t} \qquad (4\text{-}2)$$

The average velocity, $\bar{v}$, is equal to the initial velocity, v_o, plus the final velocity, v_f, divided by 2. Mathematically, this is

$$\bar{v} = \frac{v_o + v_f}{2} \qquad (4\text{-}3)$$

When this is substituted in Equation 4-2 for v, we have

$$\frac{v_o + v_f}{2} = \frac{s}{t}$$

Rearranging and solving for s, we have

$$s = \left(\frac{v_o + v_f}{2}\right)t \qquad (4\text{-}4)$$

Equations 4-2 and 4-4 are frequently used in analytical solutions of velocity problems.

In Equation 4-4 the velocity changes from an initial value, v_o, to a final value, v_f. For this to happen, there must be an acceleration. Acceleration in Chapter 2 is defined as the rate of change of velocity with respect to time. When we introduce acceleration to those cases where the velocity changes, several additional useful equations are obtained. *The acceleration in all these cases is assumed to be constant.*

From the equation $a = dv/dt$ in Chapter 2, we can obtain the following:

$$a = \frac{v_f - v_o}{t} \qquad (4\text{-}5)$$

Other useful equations can be obtained by rearranging these equations. Some of them follow.

Solving Equation 4-5 for v_f gives

$$v_f = v_o + at \qquad (4\text{-}6)$$

Substituting this for v_f in Equation 4–4 gives

$$s = v_o t + \tfrac{1}{2}at^2 \tag{4–7}$$

By similar manipulation, another equation can be developed:

$$s = \frac{v_f^2 - v_o^2}{2a} \tag{4–8}$$

We can now summarize as follows.

1. **In certain types of problems involving linear velocity, analytical solutions using the equations here are possible.**
2. **These equations have been developed making certain assumptions. They are:**
   (a) $\bar{v}$, **in Equation 4–1, is an *average* velocity.**
   (b) If acceleration exists, it must be constant.
   (c) v_f **and** v_o **are actual velocities that exist at a particular instant.**
3. **The equations are repeated here.**

$$\bar{v} = \frac{s}{t} \tag{4–2}$$

$$s = \left(\frac{v_o + v_f}{2}\right)t \tag{4–4}$$

$$a = \frac{v_f - v_o}{t} \tag{4–5}$$

$$v_f = v_o + at \tag{4–6}$$

$$s = v_o t + \tfrac{1}{2}at^2 \tag{4–7}$$

$$s = \frac{v_f^2 - v_o^2}{2a} \tag{4–8}$$

Sample problems illustrating the use of these equations are given next.

EXAMPLE
A car starts from rest and accelerates at a constant rate of $10 \, \text{ft/sec}^2$ until it reaches a velocity of 30 mph. It then travels in a straight line for 1 mile, after which it decelerates

at a constant 14 ft/sec² until it stops. Determine the time required for each segment of the travel.

Solution

A sketch of the velocity–time relationship is helpful in showing what happens to the car. This is shown in Figure 4–1. The value of the velocity is known and can be shown on a separate velocity scale. The three time intervals have been designated t_1, t_2, and t_3. The accelerations are constant at the beginning and end of the travel (note that *deceleration* is *negative* acceleration). The acceleration during the middle portion is zero. With these conditions we can use the equations that have been developed.

Equation 4–5 can be used for the first segment. The initial velocity, v_o, is equal to zero, and the final velocity, v_f, is 30 mph. The acceleration is also known. Time, t, is the unknown. The equation is rearranged as follows to solve for t directly.

$$a = \frac{v_f - v_o}{t} \qquad (4\text{–}5)$$

Simultaneously multiplying both sides by t and dividing by a, we have

$$t = \frac{v_f - v_o}{a}$$

The values can now be substituted into the equation.

$$t_1 = \frac{\dfrac{(30\,\text{miles/hr})(5280\,\text{ft/mile})}{3600\,\text{sec/hr}} - 0}{10\,\text{ft/sec}^2}$$

$$= \frac{44\,\text{ft/sec} - 0}{10\,\text{ft/sec}^2}$$

$$= 4.4\,\text{sec}$$

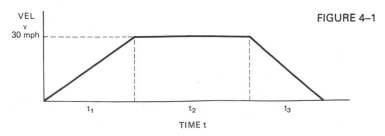

FIGURE 4–1

To be sure that the units of the problem are correct, they are carried through to the end. This practice should be followed on all problems.

Equation 4–2 can be used to solve for t_2. When rearranged, it becomes

$$t_2 = \frac{s}{v}$$

Substituting,

$$t_2 = \frac{(1\text{ mile})(5280\text{ ft/mile})}{44\text{ ft/sec}}$$

$$= 120\text{ sec} = 2\text{ min}$$

Note that, for this part of the problem, v_f in the first part is also $\bar{v}$ in the second part.

In the last part of the problem, t_3 is obtained in a similar manner as t_1. However, note that here the initial velocity, v_o, is 30 mph, or 44 ft/sec. The final velocity is zero. The equation used is

$$t_3 = \frac{v_f - v_o}{a}$$

Substituting,

$$t_3 = \frac{0 - 44\text{ ft/sec}}{-14\text{ ft/sec}^2}$$

$$= 3.14\text{ sec}$$

The negative sign for acceleration is used, since acceleration is a vector quantity and its sense must be considered.

EXAMPLE
A rock is dropped into a deep open well. Two seconds later the rock hits the water. How deep is the well?

Solution
The acceleration that acts on the rock is that of gravity, 32.2 ft/sec². Equation 4–7 is used to obtain the answer. The equation is

$$s = v_o t + \tfrac{1}{2} a t^2$$

Substituting, with $v_o = 0$, $a = 32.2 \text{ ft/sec}^2$, and $t = 2 \text{ sec}$, we have

$$s = (0)(2 \text{ sec}) + \tfrac{1}{2}(32.2 \text{ ft/sec}^2)(2 \text{ sec})^2$$

$$= 64.4 \text{ ft}$$

4–3 ANGULAR VELOCITY

Chapter 2 defined *angular velocity* as the rate of change of *angular* displacement with respect to time. This definition has similarities to the definition of linear velocity, and from this we might assume that angular velocity may have other similar relationships to angular displacement and angular acceleration. This is true, and equations relating angular velocity, angular displacement, and angular acceleration can be developed.

It was noted earlier that angular displacement can be measured in degrees, revolutions, or radians. For all equations on angular displacement, angular velocity, and angular acceleration, *radians* must be used directly or a suitable conversion to radians must be included.

To develop the equations for angular motion, consider the link that rotates about a fixed axis in Figure 4–2. It rotates at a constant angular velocity ω (omega) in radians per second or radians per minute. Its angular displacement in a time period is θ (theta). With these conditions, there is an equation relating angular velocity, angular displacement, and time that is similar to the one developed for linear velocity:

$$\omega = \frac{\theta}{t} \tag{4–9}$$

If the angular velocity changes as shown in Figure 4–3, the link now has an *angular acceleration*, which we shall call alpha (α). Under these conditions there are similar equations relating angular velocities, accelerations, displacements, and time, as there are for their corresponding linear modes.

FIGURE 4–2   FIGURE 4–3

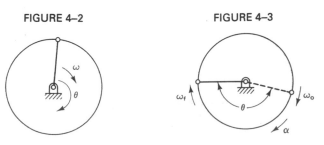

Some useful equations of this kind follow.

$$\theta = \left(\frac{\omega_o + \omega_f}{2}\right)t \qquad\qquad \textbf{(4–10)}$$

$$\alpha = \frac{\omega_f - \omega_o}{t} \qquad\qquad \textbf{(4–11)}$$

$$\omega_f = \omega_o + \alpha t \qquad\qquad \textbf{(4–12)}$$

$$\theta = \omega_o t + \tfrac{1}{2}\alpha t^2 \qquad\qquad \textbf{(4–13)}$$

$$\theta = \frac{\omega_f^2 - \omega_o^2}{2\alpha} \qquad\qquad \textbf{(4–14)}$$

The angular acceleration, α, must be constant for these equations to apply.

The similarities between these equations and Equations 4–2 and 4–4 through 4–8 are easily noticed. However, when using these equations for angular motions, there are some important differences.

1. The terms for angular velocity and angular acceleration are always in *radians.*
2. Unlike linear units, a radian in an equation is *dimensionless.* This allows proper cancellation of units in some other equations that we shall take up later.

We should also note that angular acceleration may be positive or negative, just as linear acceleration.

The examples below illustrate the use of these equations.

EXAMPLE
The flywheel on an engine is accelerated from 200 revolutions per minute (rpm) to 865 rpm in 15 sec. Find the angular acceleration and the number of turns made.

Solution
In Chapter 3 the relationship between one revolution and a radian was given in Equation 3–1. It is repeated here.

$$2\pi \text{ rad} = 360° = 1 \text{ rev} \qquad\qquad \textbf{(3–1)}$$

Revolutions per minute is the most common measurement unit used in obtaining angular velocity. However, it must be converted into radians per unit time for use in calculations, unless the conversion is directly or indirectly already incorporated into the particular equation used.

Converting the initial and final angular velocities into radians per second, we have

$$\omega_0 = \frac{(2\pi \text{ rad/rev})(200 \text{ rev/min})}{60 \text{ sec/min}}$$

$$= 20.9 \text{ rad/sec}$$

$$\omega_f = \frac{(2\pi \text{ rad/rev})(865 \text{ rev/min})}{60 \text{ sec/min}}$$

$$= 90.6 \text{ rad/sec}$$

The angular acceleration is found by using Equation 4–11.

$$\alpha = \frac{\omega_f - \omega_o}{t}$$

Substituting,

$$\alpha = \frac{90.6 \text{ rad/sec} - 20.9 \text{ rad/sec}}{15 \text{ sec}}$$

$$= 4.6 \text{ rad/sec}^2$$

Since the number of turns made is simply another method for specifying angular displacement, Equation 4–10 may be used.

$$\theta = \left(\frac{\omega_o + \omega_f}{2}\right) t$$

Substituting,

$$\theta = \left(\frac{20.9 \text{ rad/sec} + 90.6 \text{ rad/sec}}{2}\right) 15 \text{ sec}$$

$$= 836.3 \text{ rad}$$

$$\text{number of turns} = \frac{836.3 \text{ rad}}{2\pi \text{ rad/rev}} = 133.1 \text{ rev, or turns}$$

EXAMPLE
A wheel is accelerated from an initial angular velocity of 12 rad/sec to a final angular velocity of 37 rad/sec. The angular acceleration is 3 rad/sec^2. What is the angular displacement?

Solution
Equation 4–14 can be used directly for this problem. The

equation is

$$\theta = \frac{\omega_f^2 - \omega_o^2}{2\alpha}$$

Substituting,

$$\theta = \frac{(37 \text{ rad/sec})^2 - (12 \text{ rad/sec})^2}{(2)(3 \text{ rad/sec}^2)}$$

$$= 204.2 \text{ rad}$$

4–4 ANGULAR–LINEAR VELOCITY RELATIONSHIP

In studying velocities, it is logical to assume that there is a relation-ship between angular and linear velocity. This relationship is the basis that allows the linear velocity of an automobile to be measured by using the angular velocity of a rotating shaft in the transmission and showing the linear velocity in miles per hour on the speedometer.

Consider the rotating link shown in Figure 4–4. It rotates at a constant angular velocity ω about a fixed point O. The link is a solid, inelastic body, and the distance OA does not change. Point O has no linear velocity since it does not move. Point A has a velocity since it moves. *At the particular instant shown*, point A has an *instantaneous linear velocity*. Since OA is a constant length, the velocity of A can have no component along the line of action of OA. If there is no component along OA, the direction must be *perpendi-cular*, as shown in Figure 4–4. Velocity is a vector quantity, and we have shown a vector at A, v_A, to represent the velocity.

Now consider another point B, as shown in Figure 4–5, using the same link OA. It appears logical that v_B is in proportion to v_A as the length from O to B is in proportion to OA, the length from O to A. In fact, this is true. Using the principle of proportional triangles, the vector representing v_B can be drawn as shown. Thus, we see that the *radius* of rotation affects the magnitude of the velocity of point A and also point B.

FIGURE 4–4

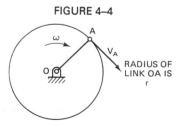

FIGURE 4–5

The angular velocity ω also affects the value of v_A. From these two considerations an important equation has been developed, as follows;

$$v = \omega r \qquad\qquad (4\text{--}15)$$

Referring again to Figure 4–4, the vector v_A is tangent to the circle at point A. For this reason, v_A is called the *tangential velocity* of the point.

The angular velocity units in Equation 4–15 again must be radians per unit of time. We have said that a radian is a dimensionless quantity. Thus, the ωr term of the equation will give units of distance/time, as required for velocity.

We can summarize as follows:

1. **The *tangential velocity* of a point moving in a circular arc is equal to the angular velocity of the link on which the point is located, multiplied by the radius of the link. The equation is**

$$v = \omega r \qquad\qquad (4\text{--}15)$$

2. **The *tangential velocity* is a vector quantity that represents the velocity of the point at a particular instance.**

One important application of this equation is brought out in the following example.

EXAMPLE

In metalworking, the term *surface feet per minute* (sfpm) means the tangential velocity of the rotating work at the point where it contacts the cutting tool. This is shown in Figure 4–6. The recommended surface feet per minute

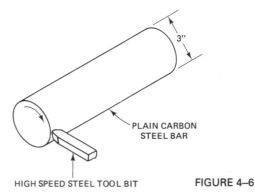

HIGH SPEED STEEL TOOL BIT   FIGURE 4–6

3″

PLAIN CARBON
STEEL BAR

depends on the material being machined and the cutting tool material. For the materials shown in Figure 4–6, the recommended velocity is 160 sfpm.

The bar is to be turned in a lathe. Determine the spindle speed of the lathe in revolutions per minute.

Solution
Application of Equation 4–15 is required for the solution. When solved for ω, the equation becomes

$$\omega = \frac{v}{r}$$

The velocity, v, is 160 ft/min. The radius is one half of 3 in., or 1.5 in. Substituting and changing to consistent units, we have

$$\omega = \frac{160 \text{ ft/min}}{1.5 \text{ in./}(12 \text{ in./ft})}$$

$$= 1280 \text{ rad/min}$$

In the preceding equation, all units except minutes cancel. The dimensionless radian is inserted into the answer for ω. The rpm, N, is obtained by converting as follows:

$$N = \frac{1280}{2\pi} = 204 \text{ rpm}$$

► **PROBLEMS**

4–1   Figure 4–7 shows the velocity–time graph of a machine element. Indicate the portions of time where the element is accelerating, decelerating, and where the acceleration is zero.

FIGURE 4–7

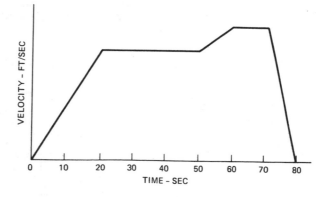

4–2 Is it possible to have a zero velocity and a high acceleration value at the same time in a mechanism?

4–3 A train accelerates from a stop to a speed of 30 mph. It takes 3 min, 20 sec to reach this speed. What is the acceleration and the distance traveled?

4–4 An automobile starts from rest and accelerates to 34 mph in 12 sec, then decelerates to 25 mph in 3 sec. It then maintains a constant 25 mph for $5\frac{1}{2}$ min. What is the total distance traveled?

4–5 A truck starts from rest and accelerates to a maximum velocity in 2310 ft. It then decelerates in 2500 ft until it stops. The total time required is 3 min. Find the maximum velocity reached and the times for the acceleration and deceleration intervals.

4–6 A small, heavy metal object is dropped from a height of 134 ft above ground. What is its velocity at the point when it hits the ground?

4–7 An airplane in flight is descending at a rate of 500 ft/min as shown on its rate of climb indicator. How long will it take to descend from an altitude of 5000 ft to 1500 ft?

4–8 A wheel accelerates from 100 to 1200 rpm in 12 sec. What is the angular acceleration?

4–9 A body is accelerated from an initial velocity of 20 ft/min to a final velocity of 86 ft/min at an acceleration rate of 2 ft/sec². What is the distance traveled?

4–10 At a deceleration rate of 3.6 rad/sec², how long will it take to stop a spindle rotating at 12,000 rpm?

4–11 It is desired to stop a wheel that rotates at 100 rpm in 1 rev. What deceleration rate is required?

4–12 A carbide cutting tool is to be used in turning a 6-in.-diameter alloy steel bar. The recommended cutting speed is 400 fpm. Find the correct spindle speed of the lathe.

4–13 A wheel rotates about an axis as shown in Figure 4–8. The angular velocity is constant. Show vectors at points *A* and *B* that represent the linear velocity of these points. Magnitude is not given and is to be neglected in your answer.

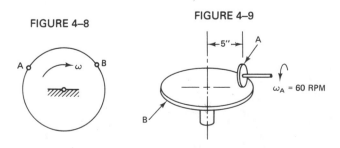

FIGURE 4–8

FIGURE 4–9

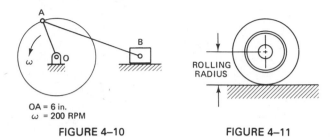

OA = 6 in.
ω = 200 RPM

FIGURE 4–10   FIGURE 4–11

4–14 Figure 4–9 shows a friction drive, with disk *A* driving disk *B*. The diameter of disk *A* is 2 in. Find the angular velocity of *B*, (*Hint:* A point on *B* in contact with the outside diameter of *A* has the same linear velocity as the point on the diameter on *A*.)

4–15 In Figure 4–10, what is the velocity of point *A*? What is the direction and sense of the velocity of the slider *B*?

4–16 An automobile wheel, as shown in Figure 4–11, has a rolling radius of 12 in. At 60 mph, what is the angular velocity of the rear axle?

4–17 Figure 4–12 shows a crank rocker mechanism. Link 1, the crank, is $6\frac{1}{2}$ in. in length, and link 3 is 12 in. in length. Find the velocities of points *A* and *B* for the position shown.

4–18 A grinding wheel has a maximum allowable velocity of 4000 fpm to ensure that wheel breakage due to centrifugal force does not occur. What is the maximum spindle revolutions per minute that can be used if the wheel diameter is 1 ft?

FIGURE 4–12

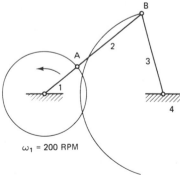

ω_1 = 200 RPM

CHAPTER 5
VELOCITY ANALYSIS

5-1 INTRODUCTION

Chapter 4 presented some useful algebraic equations for performing analytical solutions to velocity problems when conditions
permit. In this chapter we shall take up *graphical* methods that will
add to our ability to solve for velocities of points and links in a
mechanism.

5-2 RELATIVE AND ABSOLUTE VELOCITY

In Chapter 2, *relative* motion of a point or body is defined as the
motion of that point or body with respect to some other point or
body. *Absolute* motion is defined as the motion of a point or body
with respect to earth. The symbol for velocity, v, previously used,
will also be used for relative and absolute velocity. Subscripts with
this letter tell us what type of velocity we are concerned with, as in
the following examples.

Absolute Velocity: v_A

**When only one letter or subscript is used, absolute velocity is always
understood.**

Relative Velocity: $v_{A/B}$

**The subscript A/B always indicates *relative* velocity and is read as
the *relative velocity* of A with respect to B.**

Two automobiles traveling in a straight line in the same direction at
40 mph have zero relative velocity, or, $v_{A/B} = 0$ mph; however, if
one car travels in the *opposite* direction at 40 mph, the relative
velocity of the two is 80 mph, or $v_{A/B} = 80$ mph.

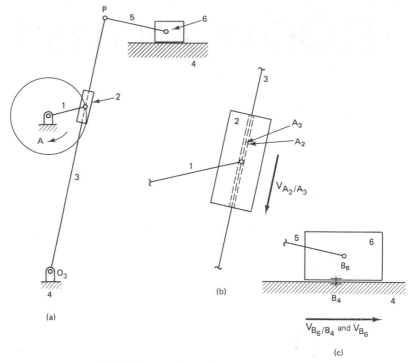

FIGURE 5–1. Quick-return mechanism.

The quick-return mechanism shown in Figure 5–1 can be used to study relative velocities. As link 1 rotates, link 2 slides up and down on link 3. Now consider a point A_2 on link 2 and in contact with link 3. The point it contacts on link 3 is A_3. Any *relative* motion and velocity of A_2 with respect to A_3 must be along the line O_3P.

Figure 5–1(b) shows an enlarged portion of the links in question, with the points A_2 and A_3. We have said velocity is a vector quantity; this statement applies to both relative and absolute velocity. Thus, we can represent v_{A_2/A_3} by a vector. In the position of the linkage shown, the vector has the direction and sense shown in Figure 5–1(b). We know nothing of the magnitude, so the length of the vector is not to scale.

Similarly, we can show a vector representing the relative velocities of a point B_6 on link 6 and a corresponding point B_4 on link 4. The relative velocity v_{B_6/B_4} has the direction and sense shown in Figure 5–1(c).

Link 4 is the fixed link of the mechanism, and all points on it have zero velocity. Thus, v_{B_4} also equals zero. With v_{B_4} equal to zero, the relative velocity of B_6 with respect to B_4 then is also the absolute velocity of B_6.

The use of vectors for absolute and relative velocities gives us a very useful method for determining unknown velocities in a mechanism. We shall demonstrate this using Figure 5–2, which is the portion of the quick-return mechanism shown in Figure 5–1(b), except that part of link 2 has been removed for clarity.

Now consider a point A located at the pin joint on link 1. This point is also on link 2, since it is located on the pin joint of link 2. Point A is thus common to links 1 and 2. The point on link 3 that is in contact with point A is called point B. It has relative motion with A, and the relative velocity is represented by the vector $v_{A/B}$. Referring to Figure 5–1(a), we know from Chapter 4 that the absolute velocity of A is a vector tangent to the circular path of point A and located at A. Its magnitude is $\omega_1 r_1$, where r_1 is the radius of the circular path (equal to the length of link 1). This vector is shown in Figure 5–2. Since we know nothing of the magnitudes of V_A and $v_{A/B}$, their lengths are not to scale.

Again referring to Figure 5–2, a vector v_B of indefinite length is drawn in perpendicular to link 3. This is the tangential velocity of point B on rotating link 3. Its magnitude is equal to $\omega_3 r_3$, where r_3 *here is the length* O_3B, not the complete length of link 3.

Vectors can be added, and we now write an equation relating them:

$$v_A = v_B + \rightarrow v_{A/B} \qquad (5\text{–}1)$$

This equation can be stated in general terms as follows:

The velocity of a point is equal to the velocity of a second point plus the velocity of the first point with respect to the second.

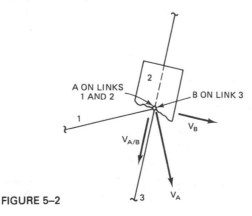

FIGURE 5–2

This is one of the most important tools we have in analyzing velocities. It applies to two points on the same body, to points on different bodies, and to all other situations. The example that follows shows how this rule may be applied.

EXAMPLE
An automobile is traveling at 50 mph. One wheel is shown in Figure 5–3. Determine the velocities of points B and C.

Solution
The center of the wheel, point O, has a velocity equal to the velocity of the automobile. The vector v_O shows this in Figure 5–3. To write equations, vectors for the relative velocities of B and C with respect to O are needed. Here the process of inversion is useful. Consider the wheel rotating as shown in Figure 5–4, with point O assumed stationary. An *inversion* has the same relative motions as the original mechanism. This means that the relative velocities of B, C, and O are the same. Since B and C rotate around O, the vectors in Figure 5–4 are the relative velocities.

Now we can write the equations. Taking first point B,

$$v_B = v_O + \rightarrow v_{B/O}$$

Substituting the vector of $+50$ mph for v_O and -50 mph for $v_{B/O}$ gives

$$v_B = +50 - 50$$

$$= 0$$

FIGURE 5–3   FIGURE 5–4

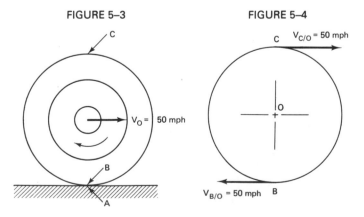

The algebraic solution is possible here only because both vectors have the same direction. If this had not been true, a graphical solution would have been used.

The fact that the velocity of *B* at this instant is zero is borne out by logic. If there is no slip between the tire and the pavement, there is no relative movement between point *B* and its contact, point *A* on the road, as shown in Figure 5–3. Point *A* is stationary, and therefore *B* also has to be stationary.

The equation for *C* is

$$v_C = v_O + \to v_{C/O}$$

Now v_O and $v_{C/O}$ are both directed to the right and therefore have positive signs. When substituted

$$v_C = +50 + 50$$
$$= 100 \text{ mph}$$

5–3 RELATIVE VELOCITY OF TWO POINTS ON THE SAME BODY

When two points are on the same rigid body, the relative velocity of the two has a *direction* that is always perpendicular to a line connecting the two points. This can be shown by examining Figure 5–5. Here is a solid, inelastic link with two points on it, *A* and *B*. Any velocity not perpendicular to a line connecting the two points will have a component parallel to the line, indicating movement of one point to or away from the other point. This is shown in Figure

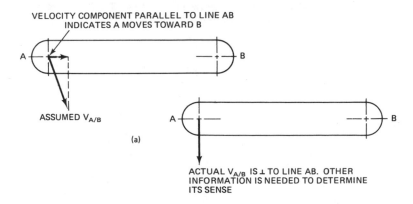

VELOCITY COMPONENT PARALLEL TO LINE AB
INDICATES A MOVES TOWARD B

A — B

ASSUMED V$_{A/B}$

(a)

A — B

ACTUAL V$_{A/B}$ IS ⊥ TO LINE AB. OTHER INFORMATION IS NEEDED TO DETERMINE ITS SENSE

FIGURE 5–5   (b)

5–5(a). This is not possible since the link is inelastic. The only possible vector is one perpendicular to the line between the two points. It should be noted, however, that other information has to be given in order for the *sense* of the vector to be determined.

5–4 VELOCITY IMAGE METHOD

The velocity image method is a graphical method for determining velocities. It depends on the absolute and relative velocity relationships brought out in this chapter. It also uses the relationship among tangential velocity, radius, and angular velocity to determine angular velocities of links in a mechanism.

It is called the *image* method because, where solid links with three or more points exist, an image of the link is formed in the velocity diagram. The image method can be explained by examining the construction in Figure 5–6, which shows a four-bar linkage with the dimensions as given and an angular velocity of link 1 of 12 rad/sec. It is desired to find the velocities of points A and B.

The principles used to arrive at a solution are those expressed by the following:

$$\mathbf{v} = \omega \mathbf{r}$$

$v_{B/A}$ **and** $v_{A/B}$ **are perpendicular to line** AB

velocity of a point = velocity of a second point $+\rightarrow$
velocity of first point/second point

FIGURE 5–6. Velocity diagram for the equation $v_B = v_A + \rightarrow v_{B/A}$.
Steps in construction:
  1. Draw v_A to scale $\perp O_1A$ and 2.3 in. long.
  2. From the terminus of v_A draw a line $\perp AB$.
  3. From the origin of v_A draw a line $\perp$ to O_3B to its intersection with $v_{B/A}$.
  4. Measure v_B to determine the velocity.

O_1A = 2.3 in.
AB = 3.7 in.
O_3B = 3.1 in.
θ = 60.5°
ω_1 = 12 rad./sec.

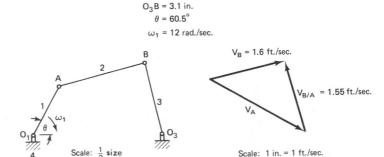

V_B = 1.6 ft./sec.

$V_{B/A}$ = 1.55 ft./sec.

V_A

Scale: $\frac{1}{2}$ size

Scale: 1 in. = 1 ft./sec.

We can find v_A as follows.

$$v_A = \omega_1 r$$

ω_1 is 12 rad/sec and

$$r = O_1 A = 2.3 \text{ in.} = \frac{2.3}{12} \text{ ft}$$

Substituting,

$$v_A = (12)\left(\frac{2.3}{12}\right) = 2.3 \text{ ft/sec}$$

We now have the magnitude of v_A and can start construction of the vector diagram. The construction steps are shown in Figure 5–6. Since v_A is the tangential velocity of the point A it is perpendicular to the line $O_1 A$.

Note that in this method it is always necessary to construct the schematic mechanism diagram to scale first, and then proceed with the velocity diagram, also to scale. Suitable scales for both have to be selected by the reader.

Arrowheads to show the correct sense are placed on the diagram as shown. The resulting vector addition represents the equation $v_B = v_A + \rightarrow v_{B/A}$, as shown in Figure 5–6. The velocity of B is determined by measuring, to the correct scale, the length of the vector v_B.

In examining Figure 5–6, it is obvious that links 1 and 3 have angular velocity since the motion is rotary. Link 2 has combined rotation and translation, and its angular velocity can be determined using the equation $v = \omega r$, where v is the *relative* velocity of any two points on the link with respect to each other, and r is the distance between the two points. We can restate this:

The angular velocity of a link is equal to the relative velocity of one point on the link with respect to a second point on the link divided by the distance between them.

The angular velocity of link 2 in Figure 5–6 is

$$\omega_2 = \frac{v_{B/A}}{AB}$$

$$= \frac{1.55}{3.7/12}$$

$$= 5 \text{ rad/sec}$$

$v_{B/A}$ was measured from the vector diagram as 1.55 ft/sec. For correct units, the dimension AB is converted to feet in the equation.

5–5 POLE POINT CONSTRUCTION OF THE VELOCITY IMAGE

The preceding section explained and illustrated how the velocity image method worked. By changing some of the methods of notation of the vector diagram, we can make it easier to use and at the same time provide additional information about the velocity types. The velocity diagram is constructed by using *pole points* for the origin of the vectors. We shall explain it by using Figure 5–7 as reference. Figure 5–7(a) shows a velocity vector diagram similar to the one in Figure 5–6; it has the same notation and the same meaning. In Figure 5–7(b), the diagram is drawn using pole point notation. The procedure for using this method follows.

Pole Point Notation

1. **All *absolute* velocities start at a pole point *o*. This point is selected first at the desired point, and the vector of the first known velocity is drawn to scale starting at point *o*.**
2. **The sense of all absolute velocities is *away* from *o*. At the end of the vector representing the absolute velocity, notate the end with the same lowercase letter used for the point on the mechanism. The notation *oa* in Figure**

FIGURE 5–7. Pole point notation of velocity diagram. With pure pole point notation, all absolute velocities start at point *o* and have their sense away from *o*. *ob* and *oa* are *absolute* velocities. All vectors connecting other lowercase letters (not *o*) are *relative* velocities. The relative velocity of *B* with respect to *A* is *from a* to *b*.

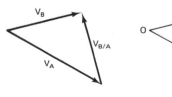

ABSOLUTE VELOCITIES ARE
V_A AND V_B.
A RELATIVE VELOCITY
IS $V_{B/A}$.

(a)   (b)

5–7(b) is read as the absolute velocity of point **A**. *Omit the arrowhead, such as used in Figure 5–7(a), and use the notation rule to indicate sense.*

3. All *relative* velocity vectors start at lowercase letters other than *o* and end at lowercase letters other than *o*. By this rule, the relative velocity of **A** with respect to **B**, or **B** with respect to **A**, must be a vector connecting points *a* and *b* in Figure 5–7(b).

4. The sense of relative velocities depends on which way the vector is read. In Figure 5–7(b), the sense of $v_{B/A}$ is from point *a* toward point *b*. Put into a rule, we can say that *the sense of the relative velocity of two points is toward the first point and away from the second point.* The arrowhead is omitted on relative velocity vectors also.

By looking at the vector notation, we can tell if a velocity is absolute or relative, since all absolute velocities start with the letter *o*. The advantages of using this method are shown in the following example.

EXAMPLE

Figure 5–8 shows a slider crank mechanism. The dimensions of the mechanism are as shown, and the velocity of

FIGURE 5–8. Steps:
1. Lay off *oa* to scale.
2. Through *o*, draw a horizontal line of indefinite length to the right.
3. From *a* draw a line $\perp AB$ until it intersects the horizontal line. The intersection is *b*.
4. From *b* draw a line $\perp BC$, length indefinite.
5. From *a* draw a line $\perp AC$ until it intersects the line in (4).
6. The intersection is *c*. Draw *oc*, the absolute velocity of *c*.
7. Measure *ob* and *oc* and the angle that *oc* makes with the horizontal.

$O_1 A$ = 7 in.
AB = 25 in.
AC = 10 in.
BC = 10 in.
V_A = 30 ft./sec.

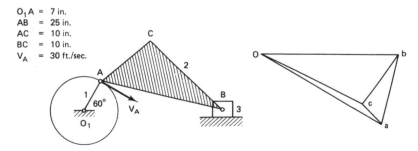

1 in. = 10 in.   1 in. = 10 ft./sec.

point A is 30 ft/sec. Determine the velocities of points B and C. Point C is on link 2.

Solution

The construction steps are shown in Figure 5–8. Points A, B, and C are three points on the same rigid body. Therefore, the relative velocity of any two of these is always perpendicular to a straight line connecting the two points. In making the construction, we are indirectly making use of the following equations:

$$v_B = v_A + \rightarrow v_{B/A}$$

$$v_C = v_B + \rightarrow v_{C/B}$$

$$v_C = v_A + \rightarrow v_{C/A}$$

Determining points b and c by intersections of vectors can be considered analogous to the algebraic method of solving two equations for two unknowns. Although this is explained in Figure 5–8, a more detailed explanation is given in Figure 5–9. Here we have used two vectors, of which we know only the direction, to add to a third vector to obtain the magnitudes of the vectors.

After the velocity diagram is complete, we now measure by scale the values of v_B and v_C. They are

$$v_B = 29.5 \text{ ft/sec, horizontal and to the right}$$

$$v_C = 24 \text{ ft/sec, to the right and down at an angle of } 25° \text{ with the horizontal}$$

FIGURE 5–9. Location of velocity points in Figure 5–8.

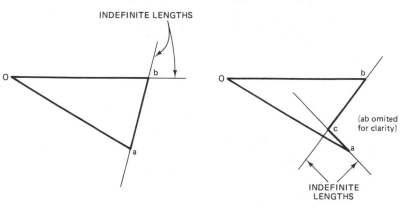

(a) Locating point b   (b) Locating point c

Note that to provide the complete answer the magnitude, direction, and sense are given. An examination of the triangle *abc* in the velocity diagram will show it is an *image* of triangle *ABC* in the mechanism schematic.

5–6 METHOD OF INSTANT CENTERS

The *method of instant centers* is also used to determine velocities of points on a mechanism graphically. It consists of finding a point common to two links that can be considered theoretically to be the instant center of rotation of the two links as they rotate *with respect to each other. Centro* is another name for instant center.

It is easy to see that the instant center of rotation of link 1 and link 2 in Figure 5–10 is the pin at O_1. Although link 2 is fixed, links 1 and 2 have relative rotary motion with respect to each other, and the pin at O_1 is common to both links. The velocity of O_1 is the same on both links and is zero.

Now consider the linkage shown in Figure 5–11. The instant center of links 1 and 3 with respect to link 4 can be seen to be at points 14 and 34. The notation 14 here will be used to show the instant center of links 1 and 4, and similarly for other links.

Links 1 and 2 are connected by a pin joint, and thus this pin is the instant center of these two links. Instant center 12 does not have zero velocity as does 14, but *relative* rotary motion between links 1 and 2 exists at any instant. Instant center 23 also exists for links 2 and 3.

Now consider the same mechanism with links 1 and 3 removed from the drawing, as shown in Figure 5–12. The two links are removed so that the relative motions of links 1 and 4 can be visualized better. It is fairly obvious that link 2 has angular velocity ω_2 as shown; however, link 4 has zero angular velocity.

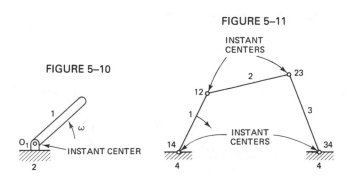

FIGURE 5–11

INSTANT CENTERS

FIGURE 5–10

INSTANT CENTER

Since relative rotary motion exists, at some point there is an instant center of rotation of link 1 with respect to link 4. This center exists even though there is no physical connection between the links involved. The point may be considered to be common to both links even though it is located beyond the physical restraints of the links.

This visualization of an instant center location of two links that can be out in space totally away from the mechanism makes understanding of the method somewhat difficult.

The *number* of possible instant centers of a mechanism is the number of possible pairs of links. It can be determined from the following equation:

$$C = \frac{n(n-1)}{2}$$ (5–2)

where

$$C = \text{number of instant centers}$$

$$n = \text{number of links}$$

The mechanism of Figure 5–11 has four links. Using Equation 5–2, the number of instant centers is

$$C = \frac{4(4-1)}{2}$$

$$= 6$$

Four of the six centers can be determined by inspection. These are the pin joints that have been identified. We now have to develop methods to locate the remaining two instant centers.

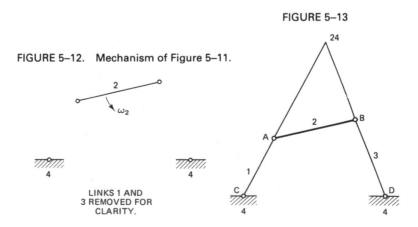

FIGURE 5–13

FIGURE 5–12.   Mechanism of Figure 5–11.

LINKS 1 AND
3 REMOVED FOR
CLARITY.

Consider the same two links, 2 and 4, of Figure 5–11. They are shown again in Figure 5–13, with points A and B on link 2 and C and D on link 4. These points also coincide with the pin joints. Now the instant center of *relative* rotational motion of links 2 and 4 is also a point common to both links. The intersection of a line drawn through A and C with another line drawn B and D gives such a point, and instant center 24 is as shown in Figure 5–13.

5–7 RULES FOR LOCATING PRIMARY INSTANT CENTERS

Instant center 24 in Figure 5–13 is called a secondary instant center because its location is found by graphical construction. The *primary* instant centers are those that can be found by inspection. We shall now list four rules for locating primary instant centers.

1. **The instant center of two links connected by a pin joint is the pin.**
2. **The instant center of two links having pure rolling contact is the point of contact between the two.**
3. **The instant center of two links having sliding contact is along a common normal at the point of contact. It can be considered to be at infinity if the surfaces are straight or at the center of curvature otherwise.**
4. **A mechanism of three links that have relative plane motion has three instant centers that lie on the same straight line. This is called the *Aronhold–Kennedy theorem*.**

Illustrations of these rules are shown in Figure 5–14. Note that the illustration of the three-link mechanism combines sliding contact of two links as well as rotational motion with pin joints.

5–8 LOCATING SECONDARY INSTANT CENTERS

An *instant center polygon* can be used to determine the remaining instant centers. In this method a regular polygon having as many sides as the mechanism has links is constructed. Connecting diagonals of the polygon show how to connect lines to locate the instant centers. The steps that follow outline the procedure.

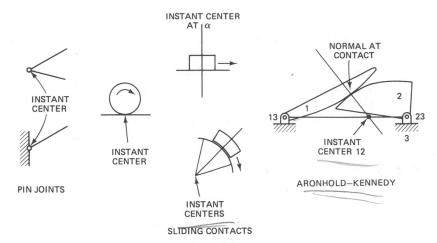

FIGURE 5–14. Examples of primary instant centers.

1. **Construct a regular polygon with the same number of sides as there are links in the mechanism.**
2. **Number each intersection of sides with a number corresponding to a link, starting with 1 and going around the polygon in one direction. This is shown in Figure 5–15(a). Each numbered side of the polygon now represents a primary instant center. For example, side 12 represents instant center 12.**
3. **Draw dashed lines connecting corners of the polygon as shown in Figure 5–15(b). These diagonals are notated by the numbers at the polygon corners. They represent the secondary instant centers.**
4. **Each dashed line adjoins four polygon sides, as shown in Figure 5–16(a) and (b). A dashed line with two polygon sides forms a triangle when the two polygon sides are on the same side of the dashed line. The three instant centers represented by the three triangle sides lie on the**

FIGURE 5–15

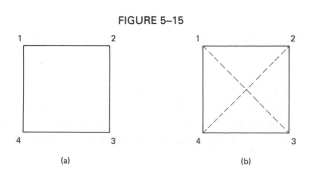

(a)   (b)

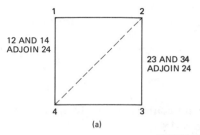

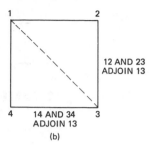

FIGURE 5–16

same line, by the Aronhold–Kennedy theorem. If we use both triangles to which the dashed line is common, the secondary instant center represented by the dashed line is located at the intersection of two lines. Thus, in Figure 5–16(a), instant center 24 is at the intersection of lines formed by connecting instant center 23 to 34 and instant center 12 to 14. Figure 5–17 illustrates this procedure in detail.

Figure 5–17 contains a method of notation that simplifies location of the correct instant centers and the line construction.

NOTE: Where the number of links is five or more, the adjoining sides of some instant centers will have one or more dashed lines in place of the solid lines shown here. This means that another secondary instant center is involved. It also means that the order in which secondary instant centers is determined is important, since there may be too many unknowns to obtain a solution.

This method is illustrated in the following problem.

FIGURE 5–17. Procedure for instant center polygon use.
Steps:
1. Construct four-sided polygon (a square).
2. Notate and draw diagonals.
3. Locate adjoining sides in the polygon, the notation 13─⎰➤12–23 ⎱➤14–34 means
   that instant center 13 is the intersection of a line drawn between instant centers 12 and 23 and a second line drawn between instant centers 14 and 34. Draw these lines.
4. Locate instant center 24 similarly. The notation 24─⎰➤12–14 ⎱➤23–34 applies in this case.

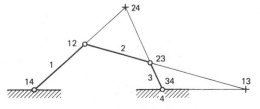

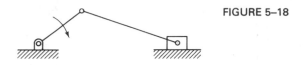

FIGURE 5–18

EXAMPLE

Find all the instant centers of the slider crank mechanism shown in Figure 5–18.

Solution

The solution is shown in Figure 5–19. The instant centers that can be found by inspection are located first and identified. They are instant centers 12, 23, 34, and 14. The polygon is next constructed as shown. The link connecting instant center 23 and 34 is simply the vertical line through 23, since 34 is a straight line of infinite length. Instant center 24 is then found as shown. The notation tells us that the line 14–34 is required to find 13. Now the only way in which instant center 34 can pass through 14 is to transfer the vertical line 34 to 14, as shown. Line 12–23 is then extended until it intersects at 13. Thus, whenever an instant center at infinity is involved, the line representing this instant center may be transferred to pass through any other instant center by drawing a line parallel to the first instant center.

5–9 USING INSTANT CENTERS TO DETERMINE VELOCITIES

The preceding sections have been devoted to ways to find instant centers. Once the instant centers have been found, we need methods to relate the instant centers to velocities.

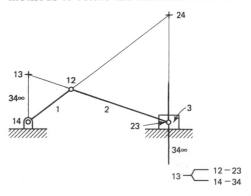

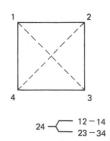

FIGURE 5–19. Instant centers for slider crank.

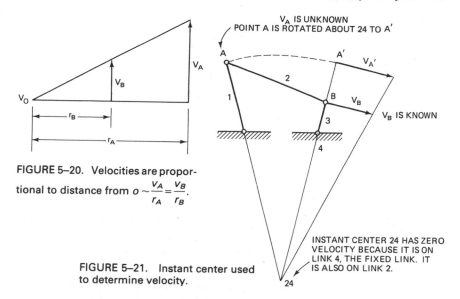

FIGURE 5–20. Velocities are proportional to distance from $o \sim \dfrac{v_A}{r_A} = \dfrac{v_B}{r_B}$.

FIGURE 5–21. Instant center used to determine velocity.

Two important characteristics allow instant centers to be used for velocity determinations:

1. **The instant center is a point common to the two bodies, or links. If the velocity of the instant center and any other point on one body are known, the velocity of any third point on the body can be found. Conversely, if the velocities of two points on a body are known, the velocity of the instant center of that body and another body can be found.**
2. **Velocities of points are proportional to their distance from the instant center and can be constructed graphically using similar triangles.**

These two characteristics are explained further in the illustrations that follow. Consider first the velocities obtained by constructing similar triangles, as shown in Figure 5–20. Velocities v_A and v_B are represented as vectors in the triangle. The velocity of point O, v_O, is zero. By the laws of similar triangles, v_A and v_B are proportional to their distances from point O, r_A and r_B, respectively. Using the equation $v = \omega r$, this also allows us to determine angular velocities of links.

Now consider the mechanism shown in Figure 5–21, which shows the construction used to find the velocity of a point by using an instant center. The velocity of another point on the same link is given. Note that the velocity of instant center 24 is zero because it is on the fixed link. This is an example of the first characteristic mentioned previously.

When point A is rotated to A', we have an example of the second characteristic using the proportionality of velocity to radius. The velocity of any other point on link 2 can also be determined in a similar manner.

We can now set up steps to use when determining velocities using instant centers. They are as follows:

1. **Note the velocity that is given in the problem and its relationship to the velocity to be determined. If the two points are on the same link, a direct solution as in Figure 5–21 can be used. If the points are on different links, an indirect solution involving another point common to both links must be used.**

2. **Select the instant center to be used, which must be an instant center for the link containing the point whose velocity is known. In Figure 5–21, instant center 24 was selected, since points A and B are on link 2. Points A and B are also on links 1 and 3, and instant center 13 could also have been used in the solution. In this case, point B, which has the known velocity, is on link 3.**

3. **Determine the velocity of the instant center. In Figure 5–21, instant center 24 has zero velocity because it is on a fixed link, but in many cases this will not be true. The alternative solution shown in Figure 5–22 is such a case; instant center 13 has a finite velocity.**

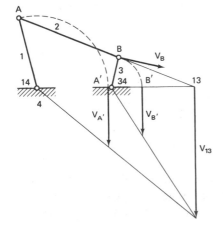

FIGURE 5–22. Alternative solution to problem in Figure 5–21. Steps:
1. Locate instant center 13.
2. Rotate point B to B' and draw $v_{B'}$.
3. Instant center 34 and B are both on link 4. Instant center 34 has zero velocity since it is also on stationary link 4.
4. Draw a line through 34 connecting the end of $v_{B'}$. Extend it until it intersects the vertical line through instant center 13. The vertical line now becomes v_{13}.
5. Connect the end of v_{13} to instant center 14.
6. Rotate point A to A' and draw $v_{A'}$ to determine the magnitude of v_A.

4. **Draw the known velocity vector in position; then draw a line connecting the ends of the instant center velocity and the known velocity. In some cases, as in Figure 5–22, the known velocity vector is rotated into position.**
5. **Rotate the point with the unknown velocity as shown in Figures 5–21 and 5–22, and draw the vector representing the desired velocity. Measure its magnitude.**

Steps similar to these are shown in Figure 5–22, and the reader is urged to study them for a better understanding of the procedure.

5–10 REVIEW

The methods for determining velocities presented in this chapter are important but also difficult to understand. The following examples are presented to help the reader gain proficiency in problem solving.

EXAMPLE
Figure 5–23 shows a quick-return mechanism. The driver, O_1A, rotates at 100 rpm. Find the velocity of point D at the instant shown.

Solution
The velocity image method is selected for the solution. The primary steps taken in the solution are listed in Figure 5–23, and additional explanation is given here.
    The velocity of point A, on the rotating driver, is calculated as

$$v_A = \omega r$$

where

$$\omega = \frac{(100 \text{ rev/min})(2\pi \text{ rad/sec})}{60 \text{ sec/min}}$$

$$= 10.5 \text{ rad/sec}$$

$$r = O_1A = 6 \text{ in.} = \tfrac{1}{2} \text{ ft}$$

Substituting,

$$v_A = (10.5)(\tfrac{1}{2}) = 5.3 \text{ ft/sec}$$

Since radian is dimensionless, the units check. The remainder of the solution involves the application of the vector equations relating absolute and relative velocities.

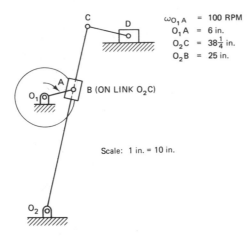

$\omega_{O_1 A}$ = 100 RPM
$O_1 A$ = 6 in.
$O_2 C$ = $38\frac{1}{4}$ in.
$O_2 B$ = 25 in.

B (ON LINK $O_2 C$)

Scale: 1 in. = 10 in.

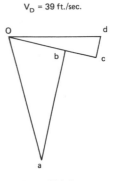

V_D = 39 ft./sec.

Scale: 1 in. = 20 ft./sec.
V_A = 5.3 ft./sec. (CALCULATED).

FIGURE 5–23. Steps:
1. Calculate v_A, using $v = \omega r$.
2. Lay off the vector $oa \perp O, A$.
3. $v_{B/A}$ is parallel to O_2C. From a draw a line parallel to O_2B of indefinite length.
4. v_B is $\perp O_2C$; from o draw $v_B \perp O_2C$, intersecting at b.
5. v_B and v_C are proportional to lengths O_2B and O_2C. On the vector diagram extend ob until the proper length oc is reached. The lengths of $38\frac{1}{4}$ and 25 in. are used to calculate relative vector lengths ob and oc.
6. v_D is horizontal since its sliding surface is horizontal. Construct ocd similarly to oab.
7. Measure v_D.

The equations used are

$$v_B = v_A + \rightarrow v_{B/A}$$

$$v_C = v_B + \rightarrow v_{C/B}$$

$$v_D = v_C + \rightarrow v_{D/C}$$

The proportionality of velocity to radius is used to determine the length oc in the vector diagram. The calculations are

$$\frac{oc}{O_2C} = \frac{ob}{O_2B}$$

$O_2C = 38\frac{1}{4}$ in. and $O_2B = 25$ in. The length of ob is now measured and found to be 24.5 ft/sec. Substituting and solving for oc, we have

$$oc = \frac{(38.25 \text{ in.})(24.5 \text{ ft/sec})}{25 \text{ in.}}$$

$$= 37.5 \text{ ft/sec}$$

The correct vector length *oc* is now laid off on the vector diagram.

We should also note that *ob* and *oc* are colinear because *B* and *C* are two points on the same body. In this special case their relative velocity is perpendicular to the line connecting the two points, and *bc* is therefore perpendicular to *BC*.

EXAMPLE

Find the velocity of point *D* in Figure 5–23 using the method of instant centers.

Solution

The solution is shown in Figure 5–24, along with the explanation and steps required to solve the problem. Only the number of instant centers necessary to find v_D have been located. The answer, 42 ft/sec, varies from that of 39 ft/sec in Figure 5–23 because of variations in the graphical construction.

EXAMPLE

The cam in Figure 5–25(a) rotates at 25 rpm. The distance from the center of rotation to the contact point on the follower is 1.75 in. at the instant shown. Find the velocity of the follower.

Solution

We shall use the velocity image method also in this problem. In Figure 5–25(b), point *P* on the cam and point *Q* on the follower represent the contact point between the two links. The radius *r* is from the center of rotation to *P*, and the velocity of point *P* must be perpendicular to *r*. Since the surface of the follower is flat and horizontal, the relative velocity of *Q* with respect to *P* is also horizontal. The velocity of the follower is vertical.

We can solve for v_P as follows:

$$v_P = \omega r$$

$$= \frac{(2\pi)(25)(1.75)}{60}$$

$$= 4.6 \text{ in/sec}$$

Units of inches were used in the answer because of the low

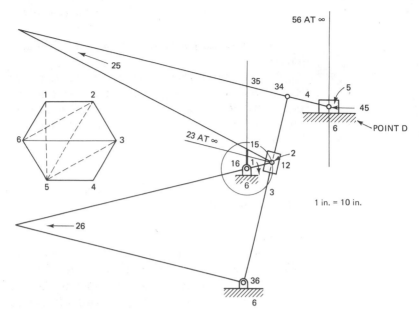

FIGURE 5–24. Method of instant centers used for a quick return mechanism. Basis ~ : Instant center 15 is on link 1 and 5 and the velocity of instant center 15 is also the velocity of link 5 (point D). With link 1 rotating about a fixed point, the velocity of any other point on it can be determined by the equation $v = \omega$, where $v = v_D$, $\omega = \omega_1$, and r = the distance 16–15.

Steps:
1. Find the instant centers necessary to locate instant center 15. Instant centers 26, 35, and 25 were found in this order, then instant center 15.
2. Measure the distance 15–16 and use the equation. Distance 15–16 is measured as 4 in.; ω_1 (from preceding problem) is 10.5 rad/sec; v_D = 42 ft/sec.

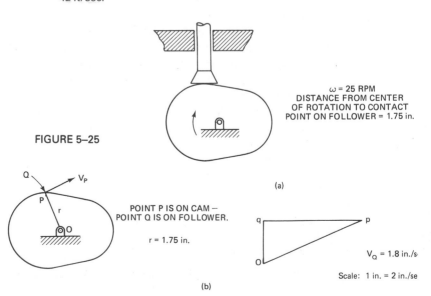

FIGURE 5–25

values, and the conversion from revolutions per minute to radians per second was made in the substitution in the equation.

The vector diagram is drawn as shown in Figure 5–25(b). The equation represented by it is $v_Q = v_P + \rightarrow v_{Q/P}$. The vector oq is measured and found to be 1.8 in./sec.

EXAMPLE

Find the velocity of the follower in Figure 5–25 using the method of instant centers. Locate all instant centers.

Solution

The cam and follower are redrawn in Figure 5–26 to show the instant centers clearly. Using Equation 5–2, the number of instant centers is

$$C = \frac{n(n-1)}{2}$$

where

$$n = \text{number of links}$$

Substituting,

$$C = \frac{3(3-1)}{2} = 3$$

Instant centers 13 and 23 are determined by inspection, using the rules previously outlined for finding primary instant centers. Links 1 and 2 have sliding contact, and the instant center for two links with sliding contact is along a common normal at the point of contact. The normal is a

FIGURE 5–26

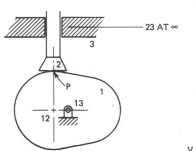
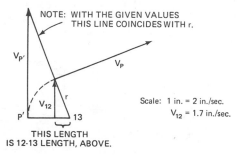

NOTE: WITH THE GIVEN VALUES
THIS LINE COINCIDES WITH r.

Scale: 1 in. = 2 in./sec.
V_{12} = 1.7 in./sec.

THIS LENGTH
IS 12-13 LENGTH, ABOVE.

VEL. OF LINK 2 = V_{12} = 1.7 in./sec.

vertical line through point *P*, as shown. Instant center 12 must also lie on the instant center line 13–23. This is a line parallel to 23 and passing through instant center 13. Thus, 12 is at the intersection of the two lines, as shown.

Because of the small number of links, the regular polygon used for finding instant centers was not constructed. If constructed, it would be a triangle with diagonal 12 and sides 23 and 13. The adjoining sides represent a line on which the instant center represented by the diagonal is located, thus justifying our location of 12.

The similar triangle construction for determining the velocity of the follower is shown below the cam in Figure 5–26. Instant center 12 is a point common to links 1 and 2. If we find its velocity, we have found the velocity of the follower also. The velocity of *P* is known from the previous example. We can then construct the triangle shown, using instant center 13 as the starting point. The vector v_{12} is located at a distance equal to the length 12–13 on the cam and away from 13. In this case, v_{12} was measured to be 1.7 in./sec. In the previous example the construction gave an answer of 1.8 in./sec. The difference in the two answers is due to the precision of the graphical construction. Enlarging the scale of the drawings would improve the precision and provide more accurate answers.

▶ **PROBLEMS**

5–1 Point *A* moves east at a velocity of 150 ft/min. Point *B* moves east at a velocity of 45 ft/min.

   (a) What is the relative velocity of *A* with respect to *B*?
   (b) What is the relative velocity of *B* with respect to *A*?

5–2 The lawn mower in Figure 5–27 is being pushed at 3 mph. Find

   (a) The velocities of *A*, *B*, and *C*.
   (b) The relative velocity of *A* with respect to *B* and *A* with respect to *C*.

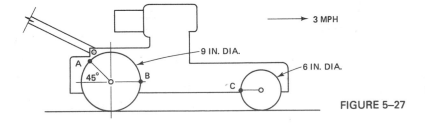

FIGURE 5–27

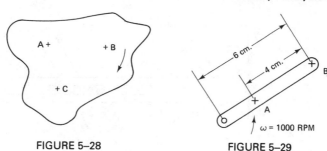

FIGURE 5–28   FIGURE 5–29

5–3 Points *A*, *B*, and *C* are on the same inelastic body, as shown in Figure 5–28. Sketch the points in the approximate same position, and then draw vectors representing the relative velocity of *A* with respect to *B*, *A* with respect to *C*, and *C* with respect to *B*. Show sense and direction only, since enough information is not given to determine magnitudes. Rotation of the body is as shown.

5–4 A link with the dimensions given in Figure 5–29 rotates at 1000 rpm. Find the relative velocity of *A* with respect to *B*.

5–5 The crank in Figure 5–30 rotates about point *O*. The velocity of point *A* is to be twice the velocity of point *B*. What should dimension *L* be?

5–6 The arm in Figure 5–31 has an instantaneous angular velocity of 100 rpm. Determine the relative velocity of the pin in the slot and the velocity of the slider.

NOTE: Problems 5–7 through 5–11, 5–16, and 5–17 are also used for acceleration problems in Chapter 6. The layouts (mechanisms and velocity polygons) done here should be retained by the student since they will be used in the acceleration solutions.

5–7 Find the velocities of points *A* and *B* in Figure 5–32, using the velocity image method.

FIGURE 5–30

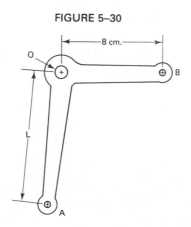

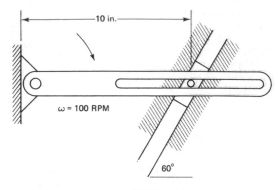

FIGURE 5–31

5–8 Figure 5–33 shows the mechanism of Figure 5–32 in a different position. Find the velocities of *A* and *B* by the velocity image method.

5–9 Find the velocity of point *B* in Figure 5–34 by the velocity image method.

5–10 The crank in Figure 5–35 rotates at 350 rpm. Find the velocities of *A*, *B*, *C*, and *D*.

5–11 Figure 5–36 shows a quick-return mechanism. Determine the velocities of *B* and *C*.

5–12 Locate all the instant centers of the mechanism in Figure 5–32.

5–13 Using the instant centers found in Problem 5–12, find the velocity of *B* in Figure 5–32.

5–14 Locate all the instant centers of the mechanism shown in Figure 5–34.

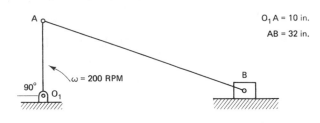

FIGURE 5–32

FIGURE 5–33

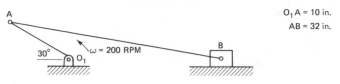

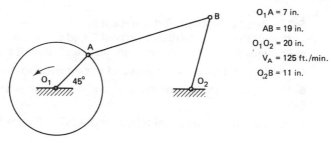

$O_1A = 7$ in.
$AB = 19$ in.
$O_1O_2 = 20$ in.
$V_A = 125$ ft./min.
$O_2B = 11$ in.

FIGURE 5–34

5–15 Determine the velocity of *B* in Figure 5–34 using the instant centers previously found.

5–16 Redraw the quick-return mechanism of Figure 5–36 with link O_1A vertical and point *A* on top. Determine the velocity of *C* in this position. Use the velocity image method.

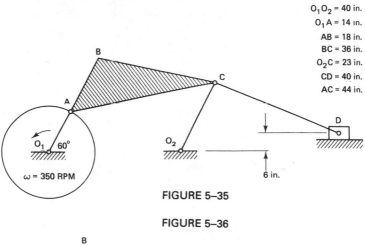

$O_1O_2 = 40$ in.
$O_1A = 14$ in.
$AB = 18$ in.
$BC = 36$ in.
$O_2C = 23$ in.
$CD = 40$ in.
$AC = 44$ in.

$\omega = 350$ RPM

6 in.

FIGURE 5–35

FIGURE 5–36

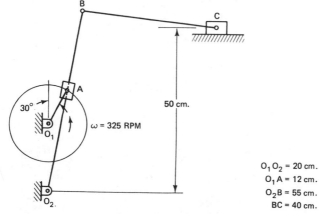

50 cm.

$30°$

$\omega = 325$ RPM

$O_1O_2 = 20$ cm.
$O_1A = 12$ cm.
$O_2B = 55$ cm.
$BC = 40$ cm.

5-17 Repeat Problem 5–16, except position point *A* at the bottom instead of the top. Find the velocity of *C* and compare with the answer in Problem 5–16. What is the significance of the difference?

5-18 Locate the instant centers of the mechanism in Figure 5–33.

5-19 Using instant centers, find the velocity of point *B* in Figure 5–33.

5-20 Find the angular velocity of link *ABC* in Figure 5–35.

CHAPTER 6
ACCELERATION

6-1 INTRODUCTION

Acceleration was defined and discussed in Chapters 2 through 4. In this chapter we consider acceleration in greater depth, and develop the relationships existing between velocity and acceleration when rotary motion is involved.

6-2 ACCELERATION COMPONENTS IN ROTATION

Points on a rotating body have at least one acceleration component at all times, even when there is no angular acceleration. This can be shown by examining Figure 6-1. Point A moves with a constant angular velocity about its center of rotation. Its linear velocity is represented by the vectors shown, each vector being perpendicular to a line from the point to the center.

Velocity is a vector quantity, and a change in magnitude or direction is a change in the vector. Thus, each time the point moves, the vector, which is the velocity, is changed. Since a change in velocity with respect to time is by definition, acceleration, there must be acceleration present. This acceleration is called the *normal* component of acceleration. It is a vector quantity from the point directed always toward the center of rotation. Its magnitude is equal to $r\omega^2$, and the symbol a^n will be used to indicate *normal acceleration*. We can restate this as follows:

A point on a rotating body has a normal acceleration that is directed away from the point toward the center of rotation. Its magnitude is equal to the radius of the rotation multiplied by the square of the angular velocity. The equation is

85

$$a^n = r\omega^2$$

(6-1)

Another useful form of this equation can be obtained by using Equation 4–15 and substituting an equivalent for ω in Equation 6–1. Equation 4–15 is

$$v = \omega r \qquad (4-15)$$

Solving for ω,

$$\omega = \frac{v}{r}$$

Substituting this term in Equation 6–1, we have

$$a^n = r\left(\frac{v}{r}\right)^2$$

Then

$$a^n = \frac{v^2}{r} \qquad (6-2)$$

Therefore, if we know the velocity of a point and its radius of rotation, we can find its normal acceleration.

A point on a body having *angular acceleration* has a *tangential* component of acceleration, as shown in Figure 6–2. It is tangent to the arc described by the point during its rotation and has the same sense as the angular acceleration. The magnitude is equal to $r\alpha$, and its symbol is a^t. The following statement summarizes.

A point on a rotating body having angular acceleration has a *tangential* acceleration a^t that is tangent to the curvature at the

FIGURE 6–1. v_a has the same magnitude at all positions, but its direction has changed.

FIGURE 6–2. Tangential acceleration of point A.

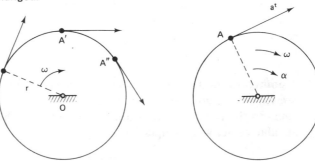

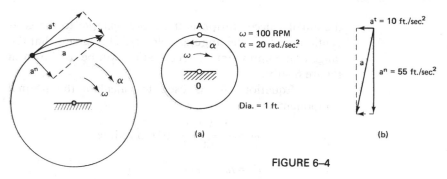

FIGURE 6–4

FIGURE 6–3

point. Its magnitude is the radius of curvature multiplied by the angular acceleration α. The equation is

$$a^t = r\alpha \qquad (6\text{--}3)$$

If a^n and a^t are vector components of the acceleration a, it follows that the acceleration a is the vector sum of the two, as is shown in Figure 6–3. In the case where there is no angular acceleration, a^t becomes zero, and a^n is the only acceleration present.

Since we are dealing with vector quantities, we can write a vector equation relating them, as follows:

$$a = a^n + \rightarrow a^t \qquad (6\text{--}4)$$

Equation 6–4 can be solved graphically by the methods we have been using. It can also be solved analytically since its form is that of a right triangle. Thus, the theorem of Pythagoras applies, and an algebraic equation can be written, as follows:

$$a = \sqrt{(a^n)^2 + (a^t)^2} \qquad (6\text{--}5)$$

The following example illustrates the application of these principles.

EXAMPLE
A wheel rotates as shown in Figure 6–4. Its angular velocity is 100 rpm and it is slowing down at the rate of 20 rad/sec^2. The wheel diameter is 1 ft. Determine the acceleration of point A at the top of the wheel.

Solution
Point A is shown in Figure 6–4(a). Note that the wheel is decelerating, and the arrowhead indicating the direction of

the angular acceleration is therefore *opposite* to the convention used for the angular velocity. This means that the tangential component of acceleration will be as shown in Figure 6–4(b).

Equation 6–1 is used to calculate the normal component.

$$\omega = \frac{(2\pi)(100)}{60} = 10.5 \text{ rad/sec}$$

$$a^n = r\omega^2 \quad \text{and} \quad r = \tfrac{1}{2} \text{ ft}$$

Substituting,

$$a^n = (\tfrac{1}{2} \text{ ft})(10.5 \text{ rad/sec})^2$$

$$= 55 \text{ ft/sec}^2$$

The tangential component is obtained by using Equation 6–3.

$$a^t = r\alpha$$

Substituting,

$$a^t = (\tfrac{1}{2} \text{ ft})(20 \text{ rad/sec}^2)$$

$$= 10 \text{ ft/sec}^2$$

Since radian is dimensionless, units of the answers come out correctly.

The acceleration a now can be obtained from Equation 6–5.

$$a = \sqrt{(a^n)^2 + (a^t)^2}$$

Substituting,

$$a = \sqrt{(55)^2 + (10)^2}$$

$$= 55.9 \text{ ft/sec}^2$$

The vector diagram showing the correct directions is drawn in Figure 6–4(b). Note that the acceleration of A is down and to the left.

6–3 RELATIVE ACCELERATION OF TWO POINTS

In Figure 6–4(a), the acceleration of point A is the absolute acceleration, since it is with respect to the fixed center of rotation O. If point O had an acceleration, we would have had a *relative* acceleration of A with respect to O. We should note that O *may* have an acceleration without being in motion. For example, the maximum acceleration of a piston occurs at the point where it stops and changes direction at top dead center.

The existence of relative and absolute accelerations means that we can vectorially add them, as we do with velocity. We can also solve graphically for acceleration using the principles of the velocity image method. We can state the rule for relating accelerations of two points in a manner similar to that used for velocity, as follows:

The acceleration of a point is equal to the acceleration of a second point plus vectorially the acceleration of the first point relative to the second point.

Consider the linkage in Figure 6–5. Points A, B, and C are related by the following equations:

$$a_A = a_B + \to a_{A/B}$$

$$a_A = a_C + \to a_{A/C}$$

$$a_B = a_C + \to a_{B/C}$$

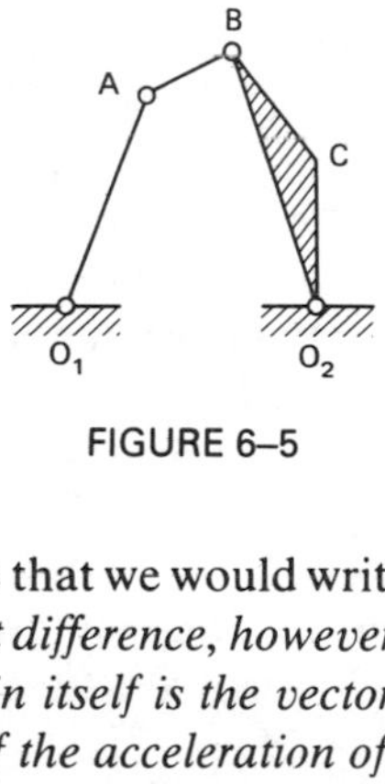

FIGURE 6–5

These equations are exactly the same that we would write if velocity were involved. *There is one important difference, however. Acceleration, a, is the total acceleration and in itself is the vector sum of the tangential and normal components of the acceleration of each point.* Thus, we have a more complicated procedure for determining acceleration than we do for velocity, since the normal and tangential components both have to be found to obtain the total acceleration.

6–4 RELATIVE ACCELERATION OF TWO POINTS ON THE SAME BODY

In Figure 6–6(a), two points A and B are on the same rigid body. Point A has an acceleration as shown. The body has a clockwise angular acceleration, α. This is the case, previously noted, when a center of rotation has acceleration. Thus, when considering the acceleration of point B, we find that we have a *relative* acceleration

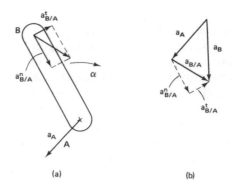

FIGURE 6–6

of *B* to *A*. This *relative* acceleration has normal and tangential components, and the vectors that represent them are shown. Here we note that their direction and sense are the same as for the point on the rotating body in Figure 6–3; thus, relative and absolute acceleration components are the same when the points are on the same rigid body.

As we noted earlier, the use of vector addition of accelerations of different points and *relative* acceleration of points on the same link allows us to find accelerations by a method similar to the velocity image method. To illustrate this, we shall use the vectors given in Figure 6–6(a) to find the acceleration of point *B*. First, writing the equation we have

$$a_B = a_A + \to a_{B/A}$$

Remembering that $a_{B/A}$ is made up of two components, we substitute these into the equation. Then,

$$a_B = a_A + \to a_{B/A}^n + \to a_{B/A}^t$$

This gives the complete form with all acceleration components. Note that we are given the total acceleration of *A*, so its normal and tangential components are not included. In many problems, however, it will be necessary to include them in the solution.

We can now proceed to draw the vectors as shown in Figure 6–6(b). The directions are the same as shown on the link in Figure 6–6(a), so the vectors in Figure 6–6(b) are drawn parallel. We have not given any magnitudes, so the lengths of the vectors are the same as in Figure 6–6(a). The resultant a_B is drawn in as the final step.

The dashed lines represent the normal and tangential components. Notice that, whether these are used without the vector $a_{B/A}$ or with it, the normal rules for head-to-tail addition of vectors applies. *The addition of these components is essentially the only thing different from the graphical solution of the velocity image method.*

6–5 ACCELERATION POLYGON

Instead of using the vectors as shown in Figure 6–6(b), the *pole point* method of construction as used in the velocity image method is a more straightforward notation and will be used here. The development of the acceleration polygon rests on the following.

1. Use of the *velocity polygon* to find needed values of velocities in order to calculate tangential and normal acceleration values.
2. Use of Equations 6–2 and 6–3 for calculating normal and tangential accelerations. The equations are

$$a^n = \frac{v^2}{r} \qquad (6\text{–}2)$$

$$a^t = r\alpha \qquad (6\text{–}3)$$

   The equations apply to and are used for finding *absolute* and *relative* accelerations.
3. A *pole point* designated as o' is the starting point for constructing the polygon. All *absolute* accelerations start at o' and are *away* from o'.
4. Lowercase letters, such as a', b', indicate the end of an absolute acceleration when prefixed with o' (for example, $o'a'$).
5. *Relative* accelerations exist when two lowercase letters are at the ends of the vector. The notation $a'b'$ is a *relative* acceleration.
6. The algebraic analogy of two unknowns and two simultaneous equations exists here also. The solution requires finding out what is known about a point and using this known in conjunction with another point to arrive at accelerations that are found by intersecting vectors.

The following example shows how to use these procedures.

EXAMPLE

The mechanism shown in Figure 6–7 rotates counterclockwise. Point A has a velocity of 8 ft/sec, and the

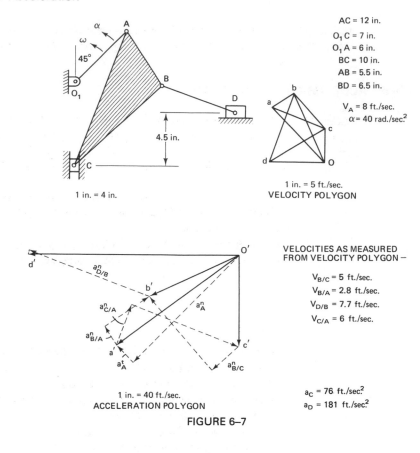

AC = 12 in.
O_1C = 7 in.
O_1A = 6 in.
BC = 10 in.
AB = 5.5 in.
BD = 6.5 in.

V_A = 8 ft./sec.
α = 40 rad./sec.²

1 in. = 4 in.

1 in. = 5 ft./sec.
VELOCITY POLYGON

VELOCITIES AS MEASURED
FROM VELOCITY POLYGON −

$V_{B/C}$ = 5 ft./sec.
$V_{B/A}$ = 2.8 ft./sec.
$V_{D/B}$ = 7.7 ft./sec.
$V_{C/A}$ = 6 ft./sec.

1 in. = 40 ft./sec.
ACCELERATION POLYGON

a_C = 76 ft./sec.²
a_D = 181 ft./sec.²

FIGURE 6–7

dimensions of the various links are as shown. Find the acceleration of points C and D.

Solution

Figure 6–7 shows the complete solution to the problem. Note that the mechanism schematic and the velocity and acceleration polygons are all drawn to scale on the same sheet. This allows drawing vectors parallel and perpendicular to the appropriate lines in the mechanism. All problems of this type done by the reader should be constructed in this manner.

Since velocity values are required to calculate normal accelerations, the first step is to construct the velocity polygon. This has been done and is shown in Figure 6–7. The values of velocity as measured are listed also in Figure 6–7.

Now we can begin construction of the acceleration polygon. The individual steps in construction are shown in Figures 6–8 through 6–11. These figures also have explanatory notes detailing the equations used and the vector construction.

Referring first to Figure 6–8, the acceleration of point A is determined as the first step. Point A is the only point that we are given any information on relating to velocity; thus, it becomes our starting point. The calculations for the normal acceleration are

$$a_A^n = \frac{(v_A)^2}{O_1A} = \frac{(8 \text{ ft/sec})^2}{6/12 \text{ ft}}$$

$$= 128 \text{ ft/sec}^2$$

Here the 6-in. link length was changed to feet to provide consistent units, and this will be done in all the calculations.
The tangential acceleration is

$$a_A^t = (O_1A)\alpha = (6/12)(4n)$$

$$= 20 \text{ ft/sec}^2$$

Starting with the pole point o', the vectors are drawn as shown in Figure 6–8.

Point C moves vertically in a straight line, and its acceleration is therefore rectilinear and along a vertical line. Points A and C are both on the same link and can be related by the vector equation $a_C = a_A + \rightarrow a_{C/A}$. Since we

FIGURE 6–8. Step 1—find a_A.

Equations: $a_A^n = \dfrac{(v_A)^2}{O,A}$

$a_A^t = (O,A)\alpha$

Vectors: a_A^n parallel to O, A
$a_A^t \perp O, A$

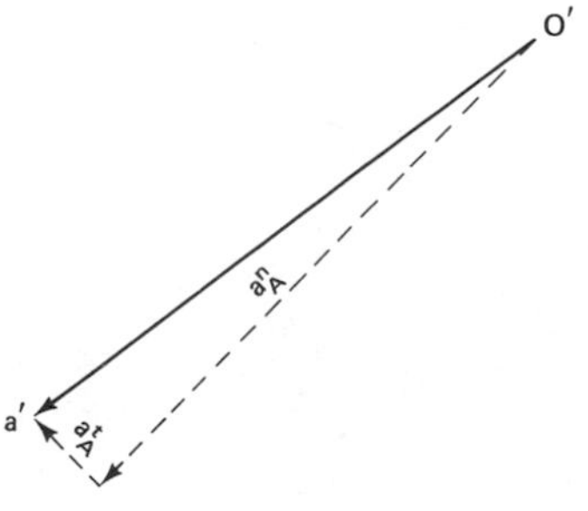

FIGURE 6–9. Step 2—find a_C.

Equations: $a_C = a_A + \rightarrow a_{C/A}$

$a_{C/A}^n = \dfrac{(v_{C/A})^2}{AC}$

Vectors: $a_{C/A}^n$ parallel to AC to required length
$a_{C/A}^t \perp AC$ and a_C vertical and intersecting $a_{C/A}^t$ at c'

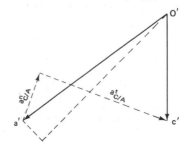

know $v_{C/A}$, the normal component can be calculated:

$$a_{C/A}^n = \frac{(v_{C/A})^2}{AC} = \frac{(6)^2}{12/12}$$

$$= 36 \text{ ft/sec}^2$$

$a_{C/A}^t$ is perpendicular to $a_{C/A}^n$ and is now drawn through the end of $a_{C/A}^n$ intersecting the vertical line through o'. $o'c'$ is measured to be 76 ft/sec, the velocity of C. The construction of this step is shown in Figure 6–9.

To find the acceleration of point D, another point on the same link must be used. The only other point on the same link with D is B. Therefore, we must determine B before we can find D. In examining the mechanism, we find that A, B, and C are all common to one link, and that the equations shown in Figure 6–10 can be written to describe the acceleration of B. The equations are

$$a_B = a_A + \rightarrow a_{B/A}$$

$$a_B = a_C + \rightarrow a_{B/C}$$

The normal acceleration values are calculated as follows:

$$a_{B/A}^n = \frac{(v_{B/A})^2}{AB} = \frac{(2.8)^2}{5.5/12}$$

$$= 17.1 \text{ ft/sec}^2$$

$$a_{B/C}^n = \frac{(v_{B/C})^2}{BC} = \frac{(5)^2}{10/12}$$

$$= 30 \text{ ft/sec}^2$$

FIGURE 6–10. Step 3—find a_B.

Equations: $a_B = a_A + \rightarrow a_{B/A}$, $a_{B/A}^n = \dfrac{(v_{B/A})^2}{AB}$

$\qquad\quad a_B = a_C + \rightarrow a_{B/C}$, $a_{B/C}^n = \dfrac{(v_{B/C})^2}{BC}$

Vectors: normal and tangential components parallel and $\perp$. Normal components are laid off to correct magnitude. Intersection of tangential components determines b'.

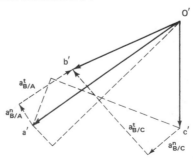

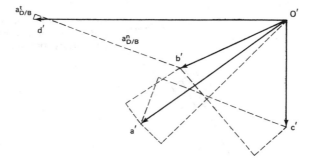

FIGURE 6–11. Step 4—find a_D.

Equations: $a_D = a_B + \to a_{D/B}$, $a^n_{D/B} = \dfrac{(v_{D/B})^2}{BD}$

Vectors: normal and tangential components parallel and $\perp$. Normal component laid off to correct length. Intersection of tangential component with horizontal line is d'.

The vectors are laid off as shown in Figure 6–10. The intersection of their tangential components locates b'. The vector $o'b'$ represents the acceleration of B.

Figure 6–11 shows the construction for finding a_D. The acceleration of D must be horizontal, since D moves in a straight line horizontally. The normal component is

$$a^n_{D/B} = \frac{(v_{D/B})^2}{BD} = \frac{(7.7)^2}{6.5/12}$$

$$= 109 \text{ ft/sec}^2$$

The acceleration of D as measured is 181 ft/sec^2.

It will be noted that tangential acceleration of D relative to B is almost zero. This would indicate that the angular velocity of link BD is almost constant at this point

6–6 ACCELERATION IN ROTATING SLIDING CONTACT LINKAGES

When a rotating link has relative sliding motion with another link, an additional acceleration component called the *Coriolis component* exists. The name of this component comes from the discoverer, Gaspard de Coriolis, a French engineer, who proved its existence in the nineteenth century. An example of this type of motion is shown in Figure 6–12, where link 2 is rotating at some angular velocity ω_2. Link 1 slides up link 2 at the same instant with a *relative* velocity $v_{1/2}$. For this type of motion, the *Coriolis* component of acceleration

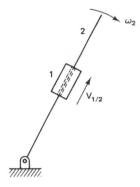

FIGURE 6–12. Relative sliding motion combined with rotary motion.

exists, as well as the normal and tangential components. The following statement summarizes the existence of Coriolis acceleration.

In a linkage that has *relative sliding* motion between links at the same time that these links have rotational motion, the *Coriolis* component of acceleration exists as well as the normal and tangential components.

The concept of the Coriolis component is difficult to develop. Because of this, we shall use a simplified explanation here. Consider the linkage of Figure 6–12 shown in the different positions of Figure 6–13(a) and (b). Link 2 rotates at a constant angular velocity. At constant angular velocity there is no angular acceleration and, therefore, no tangential components of acceleration. In Figure 6–13(a), link 1 is shown in an initial position with a relative velocity of $v_{1/2}$ along link 2. In Figure 6–13(b), link 1 has moved from its initial position to a second position. Its relative velocity $v_{1/2}$ is the same in magnitude, but *its direction has changed.* Since there is a change in the vector quantity representing velocity, *there must be an acceleration present.* This is the *Coriolis component* of acceleration.

Coriolis acceleration has both magnitude and direction. The following equation is used to determine the magnitude.

$$a^c = 2v\omega \qquad (6–6)$$

where

a^c = Coriolis acceleration
v = *relative* velocity of two points in sliding contact, one on each link
ω = angular velocity of the rotating link

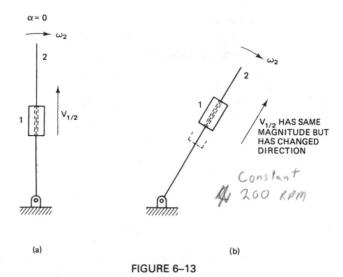

(a)   (b)

FIGURE 6–13

The *direction* of the acceleration is governed by the following statement.

The *direction* of a^c is obtained by rotating the relative velocity vector, v, 90° in the same direction as the angular velocity, ω.

Figure 6–14 illustrates this, using the same linkage as in Figure 6–12. The equation for the magnitude is also shown, using the appropriate terms for this linkage.

EXAMPLE
In Figure 6–13(a), the relative velocity of link 1 with respect to link 2 is 56 cm/sec. Link 2 rotates at a constant 250 rpm. Find the Coriolis acceleration.

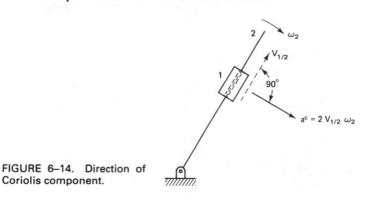

FIGURE 6–14. Direction of Coriolis component.

Solution

Equation 6–6 gives the magnitude and is used with the appropriate values substituted.

$$a^c = 2v\omega$$

where

$$v = v_{1/2} = 56 \text{ cm/sec}$$

$$\omega = \omega_2 = \frac{(200)(2\pi)}{60} \text{ rad/sec}$$

Substituting,

$$a^c = (2)(56)\frac{(200)(2\pi)}{60}$$

$$= 2346 \text{ cm/sec}^2$$

The *direction* of a^c is horizontal and to the right. This is obtained by rotating $v_{1/2}$ 90° in the same direction as ω_2.

6–7 RELATIVE ACCELERATION OF TWO POINTS AND THE CORIOLIS COMPONENT

The vector equations relating to two points A and B and their acceleration were brought out in Section 6–3. The equations stated that the acceleration of any point A is equal to the acceleration of any second point B plus vectorially the acceleration of the first point relative to the second; mathematically,

$$a_A = a_B + \to a_{A/B} \tag{6-7}$$

This was further broken down to include the tangential and normal components of acceleration that exist, as follows.

$$a_A = a_B + a_{A/B}^n + \to a_{A/B}^t \tag{6-8}$$

We have said that the Coriolis acceleration is an *additional* component if the links in question have relative sliding motion. It therefore has to be added to Equation 6–8 to make it usable for sliding motion. The equation then becomes

$$a_A = a_B + \to a_{A/B}^n + \to a_{A/B}^t + \to a^c \tag{6-9}$$

Here points A and B are the points in relative contact on the two links having sliding motion, with point A on one link and B on the other.

EXAMPLE

A cam and follower are shown in Figure 6–15. The cam rotates at 200 rpm and the distance from the center of rotation, O_1, to the follower contact point B, at the instant shown is 1.1 in. Point A is on the follower, and is the point on link 2 that contacts B on link 1. The distance from A to C, the center of the cam curvature at point B, is 0.75 in. Find the acceleration of the follower.

Solution

Since the equations used for acceleration are based on velocity, the first step is to determine the velocities of the points. This is done by constructing the velocity polygon shown in Figure 6–15. The calculated velocity of B is

$$v_B = \omega_1(O_1B)$$

$$= \frac{(2\pi)(200)}{60}\left(\frac{1.1}{12}\right)$$

$$= 1.92 \text{ ft/sec}$$

FIGURE 6–15

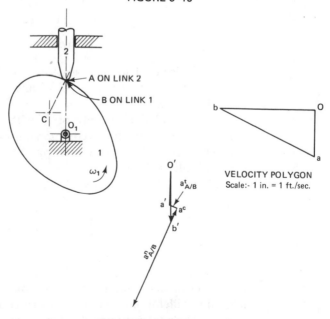

VELOCITY POLYGON
Scale:- 1 in. = 1 ft./sec.

ACCELERATION POLYGON
Scale:- 1 in. = 40 ft./sec.

Vector *ob* is laid off to scale perpendicular to the line O_1B. Note that the centerline of the cam and follower coincide and are on the same vertical line. Because the point of contact A on link 2 is on a radius, point A is slightly off and to the left of the vertical line through O_1. The effect of this is to make the direction of v_B slightly down from the horizontal; v_A is vertical, as shown. The measured values of the other velocities are

$$v_{A/B} = 2.1 \text{ ft/sec}$$

$$v_A = 0.93 \text{ ft/sec}$$

The complete equation for the acceleration of A is as follows:

$$a_A = a_B + \to a^n_{A/B} + \to a^t_{A/B} + \to a^c$$

Constant angular velocity of the cam is assumed since no angular acceleration is given. This means that there is no tangential acceleration of point B. The normal acceleration of B is

$$a^n_B = \frac{(v_B)^2}{O_1B} = \frac{(1.92)^2}{1.1/12}$$

$$= 40.2 \text{ ft/sec}^2$$

Vector $o'b'$ is now laid off to scale to start the acceleration polygon. Its direction is parallel to line O_1B, slightly off from the vertical.

The relative velocity of A with respect to B is along a perpendicular to the common normal at the point of contact. Vector ab in the velocity polygon was thus drawn perpendicular to line AC. Now the normal acceleration of A relative to B is along the normal AC. Its magnitude is

$$a^n_{A/B} = \frac{(v_{A/B})^2}{AC} = \frac{(2.1)^2}{0.75/12}$$

$$= 70.6 \text{ ft/sec}^2$$

From point b' in the acceleration polygon the vector $a^n_{A/B}$ is laid off parallel to line AC to the correct magnitude.

The next component in the equation is the tangential acceleration $a^t_{A/B}$. Its direction is perpendicular to AC, but we know nothing of its magnitude. Remembering that in vector addition the order in which the vectors are taken is unimportant, we can temporarily leave the tangential acceleration and go to the Coriolis acceleration, a^c. Its value is

$$a^c = 2(v_{A/B})(\omega_1)$$

$$= 2(2.1)\left[\frac{(2\pi)(200)}{60}\right]$$

$$= 88 \text{ ft/sec}^2$$

Its direction is the direction of $v_{A/B}$ rotated 90° counterclockwise (the same rotation as ω_1). It is drawn in from the end of $a^n_{A/B}$ up and to the right to the correct magnitude. Since the rotation of the vector makes it parallel to $a^n_{A/B}$, the addition of a^c makes both vectors colinear.

Since the follower moves vertically, its acceleration must be vertical. We also know the direction of $a^t_{A/B}$. With these directions known, we can draw the two vectors as shown in Figure 6–15. The intersection of the two is the point a'. The measured length of the vector $o'a'$ is 40 ft/sec^2, the acceleration of the follower.

▶ **PROBLEMS**

6–1 Find the normal acceleration of points A and B in Figure 6–16. The angular velocity is constant.

6–2 The wheel in Figure 6–16 accelerates at the rate of 20 rad/sec^2. The angular velocity at the instant shown is 1000 rpm. Find the tangential acceleration of A and B and the total acceleration.

FIGURE 6–16

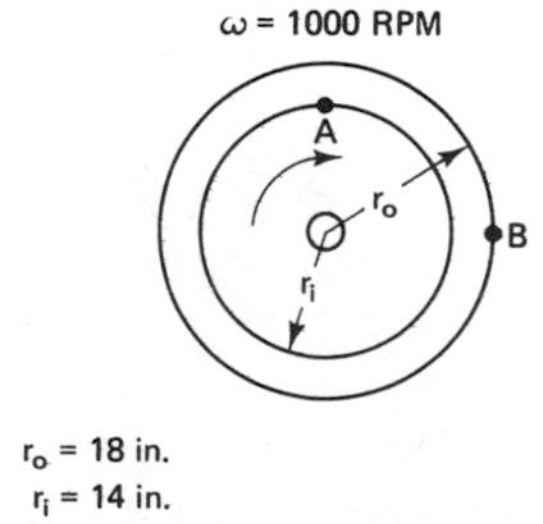

$\omega = 1000$ RPM

$r_o = 18$ in.
$r_i = 14$ in.

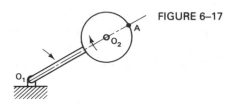

FIGURE 6–17

6–3 A 12-in.-diameter grinding wheel operating restriction specifies a maximum surface velocity of 2000 fpm to ensure that it does not disintegrate due to centrifugal force. What is the acceleration at this speed?

6–4 The grinding wheel in Problem 6–3 is operating at 1500 rpm and decelerating at 16 rad/sec². Find the total acceleration of a point on the surface.

6–5 The wheel in Figure 6–17 rotates at 100 rpm relative to the arm, while the arm rotates at 2 rpm. What is the absolute angular velocity of the wheel?

6–6 If the wheel diameter is 18 in., find the acceleration of a point on the circumference of the wheel in Figure 6–17 relative to the center of rotation O_2. Assume a constant angular velocity.

6–7 If the distance O_1O_2 is 25 in. in Figure 6–17, find the total acceleration of point *A*. The angular velocity is constant. (*Hint:* Consider points at O_2 on the arm and the wheel.)

6–8 The arm in Figure 6–18 rotates at a constant 750 rpm. Points *A* and *B* are at the distances shown. Find the acceleration of *A* relative to *B*.

6–9 If the arm in Figure 6–18 is accelerating at 20 rad/sec², what is the relative acceleration of *A* with respect to *B*? The angular velocity is 750 rpm. (*Hint:* Use the acceleration polygon, remembering that *relative* values are indicated by lowercase letters not at the pole point.)

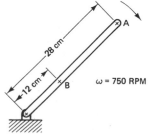

ω = 750 RPM

FIGURE 6–18

NOTE: Problems 6–10 through 6–16 use illustrations and problems from Chapter 5. The layouts and solutions for velocities obtained for Chapter 5 may be used for acceleration problems here. The acceleration polygon is to be used for solving these problems.

6–10 Find the acceleration of points *A* and *B* in Figure 5–32, Problem 5–7.

6–11 Find the acceleration of points *A* and *B* in Figure 5–33, Problem 5–8.

6–12 Find the acceleration of point *B* in Figure 5–34, Problem 5–9.

6–13 Find the acceleration of *A*, *B*, *C*, and *D* in Figure 5–35, Problem 5–10.

6–14 Find the acceleration of *B* and *C* in Figure 5–36, Problem 5–11.

6–15 Using the layout and data from Problem 5–16, find the acceleration of *C*.

6–16 Using the layout and data from Problem 5–17, find the acceleration of *C*.

6–17 Figure 6–19 shows two linkages. Does either linkage have a Coriolis component of acceleration? If so, identify it and tell why it exists.

6–18 A cam rotates at 125 rpm. The follower has a relative velocity at its point of contact with the cam of 25 ft/min, and the contact is sliding. What is the magnitude of the Coriolis acceleration?

6–19 The velocity of point *A* on link 1 in Figure 6–20 is 25 in./sec. The relative velocity of the slider on link 3 is 24.7 in./sec, and the velocity of point *B* on link 3 is 4 in./sec. Construct the acceleration diagram for points *A* and *B*, assuming constant angular velocity of link 1. (*Hint:* Write the equation for the acceleration of *A* on links 1 and 2 in terms of the acceleration of *B* and the Coriolis component. Calculate a_A^n, a_B^n, and a^c; $a_{A/B}^n$ is zero because the radius at *B* is infinity. Start the vector construction with a_A^n as the resultant and work backward in the diagram to find the unknowns.)

FIGURE 6–19

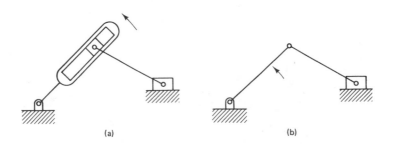

(a)   (b)

6–20 Find the acceleration of point *C* in Figure 6–20.

FIGURE 6–20

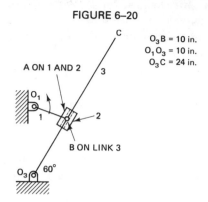

A ON 1 AND 2

O_3B = 10 in.
O_1O_3 = 10 in.
O_3C = 24 in.

B ON LINK 3

60°

CHAPTER 7
ADDITIONAL LINKAGE TYPES

7-1   INTRODUCTION

Chapter 3 introduced the reader to linkages using the four-bar linkage. Many other types of linkages have been designed to give a certain form of motion to a machine element. Although these types are too numerous to cover in this text, some of the more important ones will be covered in this chapter.

7-2   PANTOGRAPH

The pantograph is a relatively simple linkage the purpose of which is to enlarge or reduce motions traced by a point. Most commonly, it is used to enlarge or reduce drawings by tracing the outline of the drawing. It can also be adapted to cutting an enlarged or reduced part using a pattern that is traced. Oxygen cutting of metals such as steel plate frequently makes use of a pantograph for holding the oxyacetylene torch to cut the desired pattern.

The pantograph linkage is shown in Figure 7–1. To obtain the desired motion output, the linkage *CDEF* must be a *parallelogram*. In a parallelogram, opposite sides are parallel and also equal in length. Thus, *CD* equals *EF*, and *DE* equals *CF* in Figure 7–1. Pin joint *C* is fixed and the other pin joints are free. The extension of link *DE* results in one solid link *BED*, as shown. Point *B* is the tracing point that provides the enlarged replica.

If point *A* on link *EF* is selected so that *A*, *B*, and *C* are all on the same straight line, as shown in Figure 7–1, triangles *ABE* and *BCD* are similar triangles. This is true in all positions of the pantograph, since *AE* is always parallel to *CD*. The enlargement, or magnification, obtained is a ratio of the length *BC* to *AC*. In the pantograph shown, this is approximately 2.2 to 1.

The dashed lines in Figure 7–1 show the pantograph in another position. Point *A* has traced a straight line *AA'*, and the

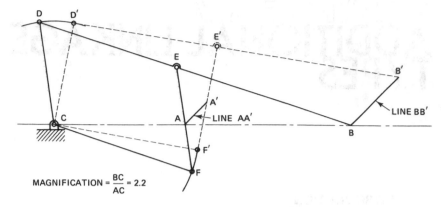

FIGURE 7–1. Pantograph linkage. Point *A* on link *EF* traces out line *AA'*. The enlarged replica of *AA'*, line *BB'*, is traced by point *B* on link *BD*.

enlarged replica has been traced by point *B*, forming another straight line *BB'* approximately 2.2 times longer than *AA'*. Although the line traced is shown straight, any path could have been followed by point *A* and its replica produced by point *B*.

Interchanging the tracing point from *A* to *B* results in a size reduction instead of enlargement. If a replica of the same size is desired, the design of Figure 7–2 is used. Here points *A*, *B*, and *F* are on the same straight line as before, but the fixed pivot point *F* is now located at the midpoint of link *DE*. This provides equal lengths *AF* and *BF*, which are necessary for exact duplication of the pattern. The path of point *A* is traced and reproduced at *BB'*, as shown. *Note that the replica formed at BB' is a mirror image of AA'.* This is not true in the pantograph of Figure 7–1.

FIGURE 7–2. Pantograph for equal-sized replica.

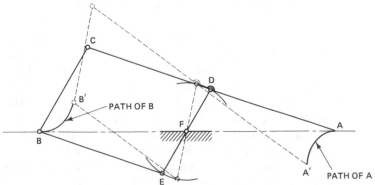

If the ratio *AF* to *BF* in Figure 7–2 is changed from the 1 to 1 ratio, we have a pantograph that provides enlargement (or reduction), just as does the pantograph in Figure 7–1. The location of point *F* on link *DE* controls this ratio.

7–3 IN-LINE SLIDER CRANK

In Figure 1–1 was shown the crank and connecting rod mechanism of a one-cylinder air-cooled gasoline engine. This is an example of the *in-line slider crank* mechanism, one of the most important and more numerous mechanism types in use. The schematic representation of the slider crank is shown in Figure 7–3 and was used in Chapter 5 to illustrate procedures for finding instant centers. Here we take up in more detail the kinematic characteristics of the mechanism.

Referring to Figure 7–3, link 1 is the crank and corresponds to the crankshaft of the engine referred to in Figure 1–1. Link 2 is the connecting rod, and link 3 represents the piston. The velocities and accelerations that exist at various displacements directly affect the *design* of the crankshaft, connecting rod, piston, and piston pin because of the forces resulting. The determination of velocity and acceleration is thus important in designing the engine.

The slider crank in Figure 7–4 is rotating at a constant angular velocity such that point *A* has a linear velocity of 40 ft/sec, as shown. At this particular instant, point *B*, the slider, has a corresponding velocity. This velocity, as well as the velocity at other positions of the crank, can be found using the velocity image method. This has been done for five crank positions, as shown in Figure 7–4(a), (b), (c), (d), and (e). The crank positions are at 45°, the position shown in the slider crank drawing, and at 0, 30, 60, and 90°. The velocity polygons are drawn for each position (the slider

FIGURE 7–3. Slider crank mechanism.

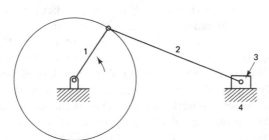

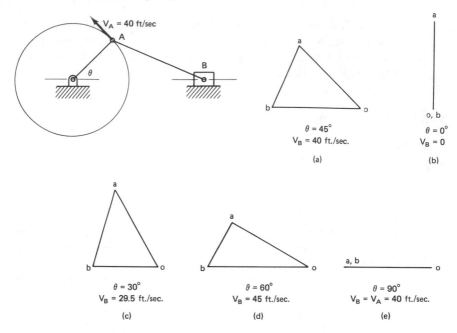

FIGURE 7–4. Slider velocity at varying crank positions.

crank itself is shown in only one position to avoid an excessive number of confusing lines).

The velocity of B is zero at the position θ equals zero, and then increases to a maximum of 45 ft/sec at 60°. At the point θ equals 90°, the velocity of B is the same as the velocity of A. Although the maximum value for the angles selected is 45 ft/sec at 60°, it is possible that a higher velocity may exist at some other angle, since the angular positions were selected arbitrarily. To determine this, we would have to select smaller increments and then graph the velocities. Before doing this, it may be wise to consider what effect the length of link AB has on our results.

In the velocity polygons of Figure 7–4, the relative velocities of A with respect to B are perpendicular to the link AB at any particular position. If we change the *length* of AB, we change also the *direction* of the relative velocity vector. The net result is a change in the velocity of B. We can show this by examining Figure 7–5. Here we show again the slider crank of Figure 7–4. Its velocity polygon is shown in Figure 7–5(a). On the drawing of the mechanism a lengthened link AB' is shown in dashed lines. Figure 7–5(b) gives the velocity polygon for this version. An examination of the vector lengths ob and ob' will show a difference in length.

The foregoing example proves that the length of the connecting link between the crank and slider has an important effect on the slider velocity. For any given design, the velocity and acceleration polygons will give the appropriate values. However, in *developing* a design, an analytical method would be faster than drawing numerous polygons. These methods exist and will be given in the following section.

7-4 EQUATIONS FOR SLIDER CRANK MOTION

The equations for slider crank motion are referenced to the symbols shown in Figure 7–6. The stroke of the slider, or piston, is of course equal to twice the radius, *r*, of the rotating crank. The displacement, *s*, of the slider is measured from the top dead center position, as shown in Figure 7–6.

The equations for *s*, the displacement, are developed in the following manner. Using Figure 7–6, the displacement *s* is equal to the distance from *B* to *O* (*OB*) subtracted from the total distance

FIGURE 7–5. Effect on velocity of changing link length.

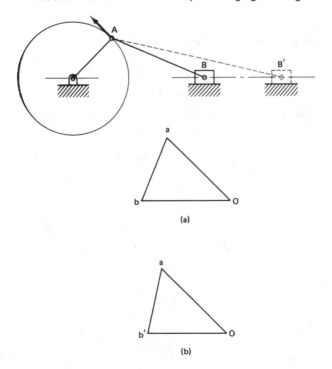

(a)

(b)

$L + r$. In equation form, this is

$$s = L + r - OB$$

where $\qquad L$ = connecting rod length

$\qquad r$ = crank radius

It now remains to express OB in terms of r, L, and the angle θ. Referring to Figure 7–6, the vertical height h can be expressed as follows:

$$h = r \sin \theta = L \sin \phi$$

Then

$$\sin \phi = \frac{r \sin \theta}{L} \qquad (7-1)$$

For convenience, we let the ratio $L/r = x$, and substitute this into Equation 7–1. Then we have

$$\sin \phi = \frac{\sin \theta}{x} \qquad (7-2)$$

Returning to the displacement s, the distance OB can be expressed as follows:

$$OB = L \cos \phi + r \cos \theta \qquad (7-3)$$

Subtracting now from the total distance $L + r$, we have

$$s = L + r - (L \cos \phi + r \cos \theta)$$
$$= r(1 - \cos \theta) + L(1 - \cos \phi) \qquad (7-4)$$

The angle ϕ can be expressed in one of the trigonometric identities,

FIGURE 7–6

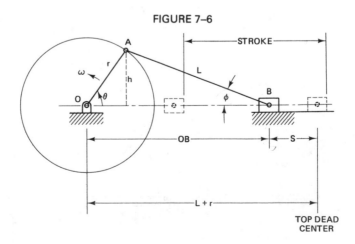

TOP DEAD
CENTER

as follows:

$$\sin^2 \phi + \cos^2 \phi = 1$$

Then,

$$\cos \phi = (1 - \sin^2 \phi)^{1/2} \qquad (7\text{--}5)$$

From Equation 7-2,

$$\sin \phi = \frac{\sin \theta}{x}$$

Substituting this for the $\sin^2 \phi$ term in Equation 7-5, we have

$$\cos \phi = \left(1 - \frac{\sin^2 \theta}{x^2}\right)^{1/2} \qquad (7\text{--}6)$$

This is now substituted in Equation 7-4 to give the following:

$$s = r(1 - \cos \theta) + L\left[1 - \left(1 - \frac{\sin^2 \theta}{x^2}\right)^{1/2}\right] \qquad (7\text{--}7)$$

Equation 7-7 is somewhat unwieldly to use, and an approximate equation has been developed that is simpler. It is

$$s = r(1 - \cos \theta) + \frac{r^2}{2L} \sin^2 \theta \qquad (7\text{--}8)$$

The equations for velocity and acceleration are derived from integrating differential equations. Because of their complicated nature, the derivations are not shown here. The *approximate* equations for slider velocity and acceleration, using point B as the locating point are as follows.

For *velocity*

$$v_B = r\omega\left(\sin \theta + \frac{\sin 2\theta}{2x}\right) \qquad (7\text{--}9)$$

For *acceleration*,

$$a_B = r\omega^2\left(\cos \theta + \frac{\cos 2\theta}{x}\right) \qquad (7\text{--}10)$$

In both of these equations, ω is the angular velocity of the crank and is constant.

Both of the preceding equations contain x, the ratio of L/r, the connecting rod length to crank radius. Also in both equations, for any given crank radius r and angular velocity ω, the value of the velocity and acceleration is a function of the terms in the parentheses. In selecting a ratio of L to r, a general knowledge of how the ratio affects the velocity and acceleration would be helpful. This can be obtained by plotting a family of curves for various values of L/r that relate the value of the angle θ to the function contained in the parentheses.

This has been done in Figures 7–7 and 7–8. Figure 7–7 plots the parenthesis term $(\sin \theta + \sin 2\theta/2x)$ against the angle θ. Four curves are plotted, for L/r values of 1.5, 2, 3, and 4. The selection of these values is made by consideration of the possible range of practical values of L/r. Practically, L must be greater than r to provide space for the crankcase and cylinder in a mechanism such as the internal combustion engine. The minimum value of 1.5 is therefore selected for the first curve and the value of 4 for the maximum. The curves indicate that lower velocities are obtained when the ratio L/r is greater; in other words, lower velocities result when the connecting rod length is greater with respect to the crank radius. An examination of the *maximum* velocities shows that there is little difference in these, however.

FIGURE 7–7

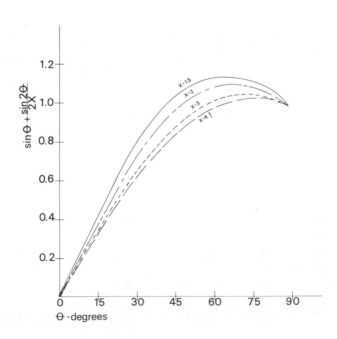

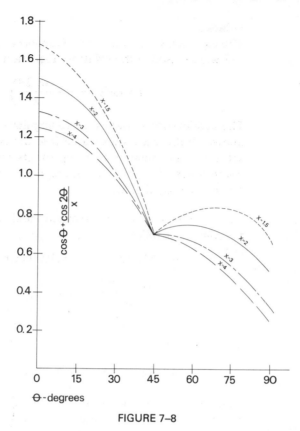

FIGURE 7–8

Figure 7–8 shows the accompanying graph for acceleration. The parenthesis term plotted is now $(\cos\theta + \cos 2\theta/x)$. Its family of curves shows a pronounced difference in accelerations using different L/r ratios. Not only are the average accelerations lower with a higher L/r ratio, but maximum acceleration values are significantly lower also.

The following example illustrates how these data are used.

EXAMPLE

It is desired to design a small single-cylinder gasoline engine. The design speed is 4000 rpm, and the crankshaft radius has been fixed at 2.5 cm. Maximum stress considerations in the piston and piston pin dictate that the maximum allowable acceleration of the piston be 6400 m/sec^2. Determine a suitable connecting rod length.

Solution

The equation for acceleration, Equation 7–10, and Figure 7–8 will be used in the solution. Equation 7–10 is

$$a = r\omega^2\left(\cos\theta + \frac{\cos 2\theta}{x}\right)$$

The acceleration a, radius r, and angular velocity ω are all known. If these are substituted into the equation, we can solve for the number that represents the terms in the parentheses. Reference to Figure 7–8 then will tell which L/r curve should be used.

The units to be substituted should be consistent, so we convert as follows:

$$a = 6400 \text{ m/sec}^2 = 640,000 \text{ cm/sec}^2$$

$$\omega = \frac{(2\pi)(4000)}{60} \text{ rad/sec}^2$$

$$= 419 \text{ rad/sec}^2$$

Substituting,

$$640,000 = (2.5)(419)^2\left(\cos\theta + \frac{\cos 2\theta}{x}\right)$$

Transposing both sides of the equation and dividing by the number before the parentheses, we have

$$\left(\cos\theta + \frac{\cos 2\theta}{x}\right) = \frac{640,000}{(2.5)(419)^2}$$

$$= 1.46$$

Referring to Figure 7–8, at $0°$, the position of highest acceleration, the factor 1.46 occurs between the curves where $x = 2$ and $x = 3$. Selecting the highest of these,

$$x = \frac{L}{r} = 3$$

Solving for L,

$$L = 3r = 3(2.5) = 7.5 \text{ cm}$$

A more careful calculation would show that the 1.46 factor could be met with a shorter length. However, it is always desirable in design to leave a reasonable factor of safety. Lengths greater than 7.5 cm would provide still lower acceleration, of course.

7-5 QUICK RETURN MECHANISMS

As the name implies, *quick-return mechanisms* are designed so that one link has slow motion in one direction of travel and fast motion in the opposite. Frequently, the link is a reciprocating slider, although an oscillating link may also be used. The shaper, a machine tool used for planing metals, is a good example of an application of the quick-return mechanism. In this case, the feed of the cutting tool is slow on its forward travel. Since there is no cutting done on the return stroke, this stroke is faster to save time. The principle is illustrated in Figure 7-9.

The shaper quick-return mechanism is a combination of the slider crank and a sliding block mechanism. It was discussed briefly in Section 5-10, and used as an example for finding velocity graphically. Here we shall examine the principle that is responsible for the quick return. Figure 7-10 shows the schematic for the quick-return mechanism. The crank, link 1, rotates at a constant angular velocity. Link 2 slides along link 3 and causes link 3 to rotate between the extreme positions shown. The other slider, link 5, is connected to the ram of the shaper. Referring to Figure 7-10, with the direction of rotation shown, link 2 moves from point A_1 to point A_2. The angle it moves through is 2θ. At the same time, link 5 moves to the right to its extreme position. Now as link 2 continues to move along the circular path back to point A_1, the angle it moves through is $360° - 2\theta$. At the same time, link 5 has returned to its other extreme position on the left. With constant angular velocity, the time required for link 2 to move through an angle is proportional to the angle. The ratio $(360 - 2\theta)/2$ now becomes the ratio between the forward and return strokes of link 5 of the shaper. Reduced to lowest terms, the ratio is $(180 - \theta)/\theta$. The angle θ, of course, depends on the length of the crank, link 1, and the distance O_1O_3.

FIGURE 7-9. Shaper with quick-return mechanism.

SLOW FORWARD FOR CUTTING

FASTER RETURN

RAM

CUTTING TOOL

WORK

SHAPER BED

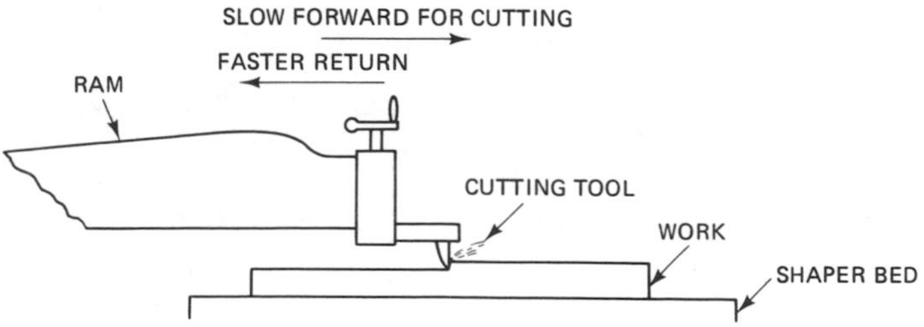

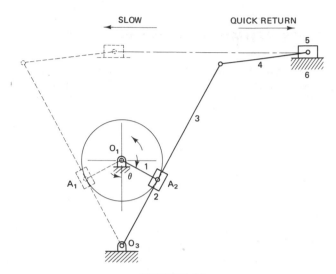

FIGURE 7–10

The *drag link* may also be adapted to a quick-return mechanism. This type is shown in Figure 7–11. The slider, link 5, is attached to the drag link by link 4. It reciprocates with the stroke as shown. The extreme positions of the slider and the corresponding positions of link 1 are shown in dashed lines. The crank, link 1, rotates at a constant angular velocity. Under these conditions, the ratio $(360 - \theta)/\theta$ is the time ratio of the forward and return strokes.

Design of the drag-link type uses the criteria of Chapter 3 for the drag-link portion with adjustment to obtain the desired ratio

FIGURE 7–11.   Drag link adapted for quick-return mechanism.

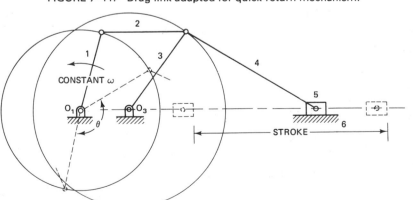

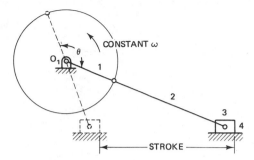

FIGURE 7–12.   Offset slider crank mechanism.

of forward and return stroke times. The design frequently requires trial-and-error techniques.

The *offset slider crank mechanism* shown in Figure 7–12 is another type of quick-return mechanism. Here the direction of slider motion is offset at an angle to the crank center rotation. The time ratio of forward and return strokes again is a function of the angle θ as long as the crank angular velocity is constant. The mechanism in Figure 7–12 is drawn in the two extreme, or limiting, positions. The time ratio of forward to return stroke is $(360 - \theta)/\theta$.

7–6   TOGGLE MECHANISM

Applications for the *toggle mechanism* are somewhat different than for other mechanisms. The toggle mechanism uses generally low velocity movements, which, due to the design of the links, can provide high forces for holding or clamping. In this case the distinction that kinematics is not directly concerned with force does not apply. The design of the toggle in Figure 7–13 is aimed at providing a high clamping or crushing force at the end of the slider. The crank rotates as shown, and the torque forces link 2 down into the extreme position indicated by the dashed lines. The small additional movement of the slider, indicated by the dashed line, provides a high force at this point. This type of mechanism is frequently used in jaw crushers for crushing ore and rock.

7–7   SCOTCH YOKE

Figure 7–14 shows the Scotch yoke. The crank rotates at a constant angular velocity, and the slider causes the yoke to reciprocate. The Scotch yoke is interesting because the yoke reciprocates with *simple*

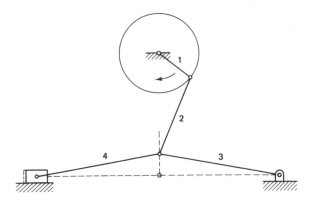

FIGURE 7–13.   Toggle mechanism.

harmonic motion. We may define simple harmonic motion as follows:

> **Simple harmonic motion.   The motion of a point moving in a straight line whose acceleration is always proportional to its distance from the starting point and always directed toward the starting point.**

Simple harmonic motion is used in cam design, and the equations for simple harmonic motion will be developed in Chapter 10. Because of wear occurring at the slider and yoke contact area, applications for the Scotch yoke are limited.

FIGURE 7–14.   Scotch yoke.

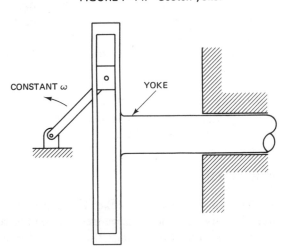

► PROBLEMS

7–1 Design a pantograph that will be used to enlarge drawings to double size. All the drawings are on $8\frac{1}{2}$- by 11-in. sheets. The pantograph design should be confined to determining the lengths of the links and to showing a schematic of the pantograph.

7–2 The pantograph in Figure 7–15 has the dimensions shown. Point *A* is the point used for tracing the pattern or drawing. The enlarged replica is reproduced at point *B*. Determine (a) the proper location of *A* on link *CD*, and (b) the magnification of the pantograph. (*Hint:* Consider as similar triangles and solve by trigonometry.)

7–3 Design a pantograph to produce a reverse, or mirror, image of the curve in Figure 7–16. The envelope dimensions of the curve are shown, and the image is to be the same size.

7–4 An in-line slider crank mechanism has a crank radius of 5 in. and a connecting rod length of 10 in. Find the acceleration of the slider when the crank angle is 45° and the speed is 1500 rpm.

7–5 Find the velocity of the slider in Problem 7–4 when the crank angle is 65°.

7–6 Find the maximum velocity of the slider in Problem 7–4. (*Hint:* Use Figure 7–7 as an aid in angle determination.)

7–7 Find the maximum acceleration of the slider in Problem 7–4.

7–8 The crank of the slider crank in Figure 7–17 is 5 in. long. At a crank angle of 75°, the displacement *s* is required to be 4.5 in. Find the length of the connecting rod *AB*.

7–9 The piston of the slider crank in Figure 7–17 is required to have a velocity of 20 in./sec at the position shown. What angular velocity of the crank, O_1A, must be used to obtain this? (The connecting rod length is that determined in Problem 7–8.)

7–10 A single-cylinder piston air compressor is to be designed with a stroke of 8 in. The maximum acceleration of the piston is 4800 ft/sec². Determine the connecting rod length if the design speed is 1000 rpm.

FIGURE 7–15

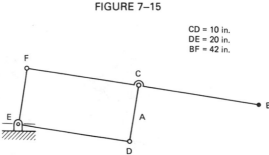

CD = 10 in.
DE = 20 in.
BF = 42 in.

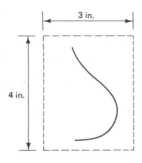

FIGURE 7–16

7–11 A slider crank has a crank radius of 10 cm and a connecting rod length of 21 cm. It rotates at a constant 250 rpm. Determine the displacement and velocity of the slider at a crank angle of 55°.

7–12 The quick-return mechanism in Figure 7–18 is to be designed so that the forward stroke of the slider takes twice as long as the return. The crank radius is 6 in. Find the correct distance O_1O_2 between the fixed centers.

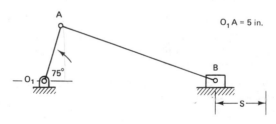

$O_1 A = 5$ in.

FIGURE 7–17

FIGURE 7–18

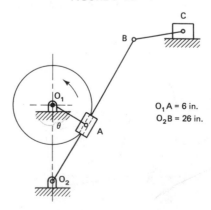

$O_1 A = 6$ in.
$O_2 B = 26$ in.

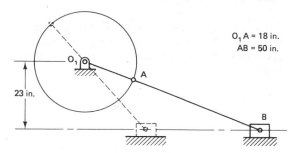

O_1A = 18 in.
AB = 50 in.

23 in.

FIGURE 7–19

7–13 Find the length of O_1O_2 in Figure 7–18 if the time ratio of forward to return stroke is 1.5 to 1. What effect does this change have on the stroke of the slider, assuming that length O_2B stays constant?

7–14 What is the length O_1O_2 in Figure 7–18 when the time for the forward stroke is the same as for the return stroke?

7–15 The offset slider crank in Figure 7–19 is used as a quick-return mechanism. It has the dimensions shown, and the slider is shown in its two extreme positions. Calculate the time ratio of its forward to return stroke.

7–16 Determine the time ratio of forward to return strokes if the vertical height of 23 in. in Figure 7–19 is changed to 20 in.

7–17 The crank of the toggle mechanism in Figure 7–20 is at the 45° position shown. When the crank moves 45° more to the vertical position, the slider moves the distance x as shown on the drawing. Find the distance x. (*Hint:* Sketch the mechanism in both positions; then set up triangles and find x as the difference of two dimensions.)

FIGURE 7–20

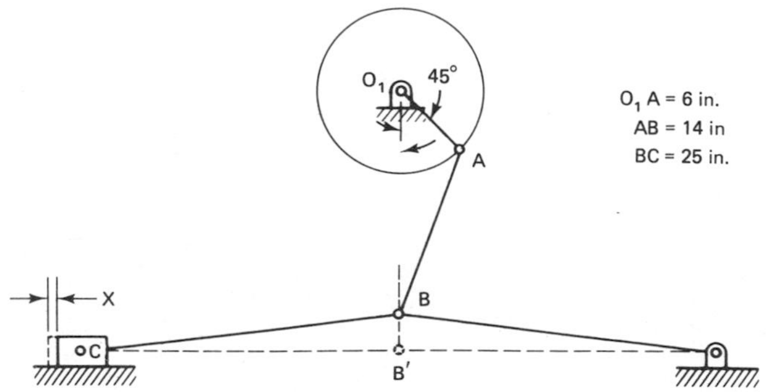

O_1A = 6 in.
AB = 14 in
BC = 25 in.

45°

7–18 In Figure 7–21, point *A* has an acceleration of 45 m/sec²
directed to the left of the starting point *O*. Position *A'* is one half the
distance to *O*, as was the original position *A*. What is the acceleration
at *A'*?

FIGURE 7–21

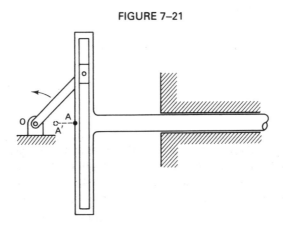

CHAPTER 8
INTERMITTENT MOTION MECHANISMS

8–1 INTRODUCTION

The motion of most of the mechanisms we have studied is *continuous* motion. That is, the mechanism links move continuously in a defined pattern frequently driven by a constant velocity crank. The *intermittent motion mechanism* is designed to provide stop-and-start motion at regular or irregular intervals. This stop-and-start motion is called *intermittent motion*.

In this chapter we shall survey some of the more important devices used for intermittent motion. Some of these mechanisms are the ratchet, the Geneva mechanism, the escapement, the star wheel, cams, and gears with removed teeth. Cams and cam design are covered separately in Chapters 9 and 10 because of their importance.

8–2 RATCHET

The *ratchet* is one of the simplest of the intermittent motion mechanisms. It is also one of the most used, in applications from hand tools, such as the ratchet wrench shown in Figure 8–1, to

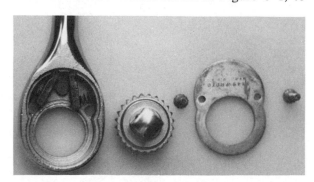

FIGURE 8–1.
Mechanism of
ratchet wrench.

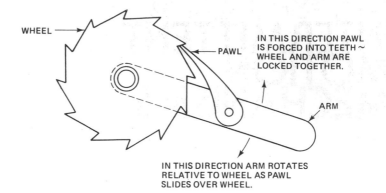

WHEEL

PAWL

IN THIS DIRECTION PAWL
IS FORCED INTO TEETH ~
WHEEL AND ARM ARE
LOCKED TOGETHER.

ARM

IN THIS DIRECTION ARM ROTATES
RELATIVE TO WHEEL AS PAWL
SLIDES OVER WHEEL.

FIGURE 8–2.   Ratchet wheel and pawl.

automobile parking-brake mechanisms and many others. The action of the ratchet can be examined in Figure 8–2, which shows the *ratchet wheel* and the connecting *pawl*. The pawl is spring-loaded so that it is always in contact with the wheel. As the arm rotates clockwise, relative motion exists between it and the wheel. When the arm rotates counterclockwise, the pawl is forced into a tooth and no relative motion can occur.

FIGURE 8–3.   Double-action ratchet. Wheel indexes in a counter-clockwise direction each time that the arm is actuated. The wheel direction is counterclockwise regardless of whether the arm is on its forward or return stroke.

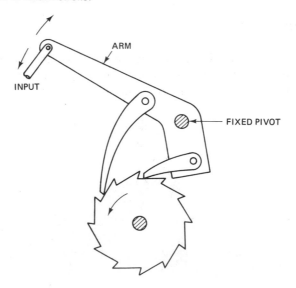

INPUT

ARM

FIXED PIVOT

Another way to use a ratchet is shown in Figure 8–3. Here the arm is used to index the ratchet wheel. Rotation of the wheel is always in the same direction regardless of arm direction of rotation.

The other ratchet designs in use are far too numerous to cover here. Some of them are used in typewriters, vending machines, counting mechanisms, stepping switches, and clocks.

8–3 GENEVA MECHANISM

Geneva mechanisms are used for indexing and providing a *dwell* during a part of the operating cycle of a machine. They are used widely on machine tools because of this ability to dwell. The term *dwell* simply means that the mechanism stays in one position for a specified time, after which it indexes to the next position.

The Geneva mechanism and its operating principle are shown in Figure 8–4. With the driver rotating at a constant angular velocity, the drive pin enters one slot, as shown at the left in Figure 8–4. The drive pin then rotates the slotted wheel until it leaves the slot. The slotted wheel stays stationary until the driver completes the revolution and reenters the next slot.

The Geneva mechanism in Figure 8–4 has four slots and, because of the geometry necessary for proper functioning, it dwells for three fourths of the driver's one revolution. The number of dwells possible from Geneva mechanisms ranges from a minimum of 3 to approximately 18 for every revolution of the slotted wheel.

In designing a Geneva mechanism, it is important that the motion of the pin at the time of entry into the slot be parallel and in

FIGURE 8–4. Geneva mechanism.

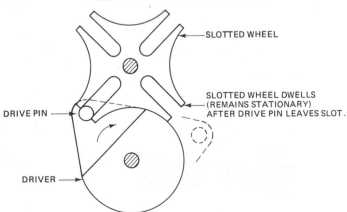

the same direction as the slot sides. This is shown in the five-station mechanism of Figure 8–5(a). In examining the geometry necessary to accomplish this, it is apparent that driver link AO_2 must make an angle of 90° with the slot. In Figure 8–5(b) the triangle AO_1O_2 is drawn showing the 90° angle. If we let 2θ equal the number of degrees between stations, the following mathematical relationships can be set up.

Letting n equal the number of stations or slots,

$$2\theta = \frac{360°}{n}$$

and

$$\theta = \frac{180}{n} \qquad (8–1)$$

The radius AO_2 of the driver then is

$$AO_2 = (O_1O_2) \sin \frac{180}{n} \qquad (8–2)$$

As the number n of slots increases, the term $180/n$ decreases. Since the sine of an angle decreases as the angle decreases, the radius AO_2 decreases also. *Thus, as the number of slots increases, the size of the driver in relation to the slotted link becomes smaller.*

FIGURE 8–5. Five-station Geneva mechanism.

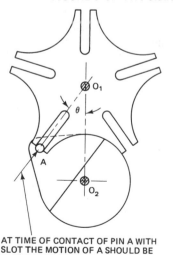

AT TIME OF CONTACT OF PIN A WITH
SLOT THE MOTION OF A SHOULD BE
IN THE SAME DIRECTION AS THE SLOT
AND MAKE AN ANGLE OF 0° WITH THE
SLOT.

$$AO_2 = (O_1O_2) \sin \frac{180°}{n}$$

(a)   (b)

We noted earlier that the slotted member of the Geneva mechanism in Figure 8–4 dwells for three fourths of the driver's revolution. Applying the triangle of Figure 8–5(b) for our analysis, when the number of stations is four, the angle θ is 180/4, or 45°. The angle ϕ in this case is also 45°, as shown by the following.

EXAMPLE

The angle at A in Figure 8–5(b) is always 90° (this is required for the pin to enter the slot parallel to the slot). The sum of the three angles in a triangle is 180°; then

$$\theta + \phi + 90° = 180°$$

$$\theta + \phi = 180° - 90° \qquad (8–3)$$

$$= 90°$$

and

$$\phi = 90° - \theta$$

Thus, when θ is 45°, ϕ is also 45°.

Now consider the five-station Geneva of Figure 8–5(a). Here the angle is

$$\theta = \frac{180}{n} = \frac{180}{5}$$

$$= 36°$$

and

$$\phi = 90° - \theta$$

$$= 90° - 36°$$

$$= 54°$$

Thus, in this case the pin is required to move 54° while the slot moves only 36°. The total movement of slot and pin is twice these values, or 108° for the driver and 72° for the slotted member.

The slotted member dwells while the driver moves through 360° − 108°, or 252°. Using the triangle and the angles of Figure 8-5(b), an equation for the *percent of dwell* for one revolution of the driver is

$$\text{percent (\%) of dwell} = \left(\frac{360° - 2\phi}{360°} \right) 100\% \qquad (8–4)$$

Equation 8–4 applies only to the *external* Geneva mechanism.

The mechanism in Figure 8–4 is an *external* Geneva mechanism. *Internal* Geneva mechanisms also exist. Such a mechanism is shown in Figure 8–6. Here we have a four-station internal Geneva mechanism with the slotted member driven by the smaller-diameter driver. Note that in this case the driver moves the driven member for most of one revolution. In fact, the driven member dwells for only one quarter of a revolution. This is a fundamental functional difference from the external Geneva mechanism. We can state it as follows:

In the *internal* Geneva mechanism the dwell period is always *less* than the motion period; in the *external* Geneva mechanism the dwell period is always *more* than the motion period.

Just as in the external Geneva, the diameter of the driver decreases as the number of slots increases.

8–4 ESCAPEMENT

The *escapement* is an intermittent motion mechanism that most of us have seen in the mechanisms of watches and clocks. Its applications, however, extend beyond watches and clocks to many other machines. The escapement can be described as a mechanism for the controlled, incremental release of stored energy. This release of energy results in finite motions of the other links in the mechanism. We shall illustrate this by describing how the escapement of the watch works.

Figure 8–7 shows a typical watch escapement with its primary parts. These are the *escape wheel*, the pallets, and the *pallet lever*. In the position shown in Figure 8–7, the upper pallet restrains the escape wheel so that it cannot turn. The motion of the pallet lever is oscillatory; that is, it moves back and forth as the balance

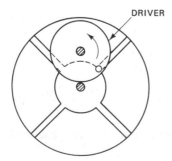

DRIVER

FIGURE 8–6. Internal Geneva mechanism.

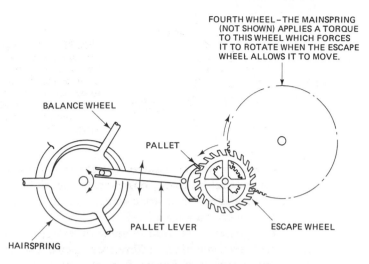

FOURTH WHEEL – THE MAINSPRING
(NOT SHOWN) APPLIES A TORQUE
TO THIS WHEEL WHICH FORCES
IT TO ROTATE WHEN THE ESCAPE
WHEEL ALLOWS IT TO MOVE.

BALANCE WHEEL

PALLET

PALLET LEVER

ESCAPE WHEEL

HAIRSPRING

FIGURE 8–7.   Watch escapement.

wheel oscillates. At the other extreme of its motion, the lower pallet
restrains the escape wheel. In between these extremes, nothing
restrains the escape wheel and it *escapes* until stopped by the next
pallet. The fourth wheel is geared to the escape wheel and thus
moves with it. The fourth wheel is the wheel carrying the second
hand; it meshes with a wheel called the third wheel (not shown in
Figure 8–7), which carries the minute hand.

To complete this description of escapement functioning,
we now have to consider the energy input and forces applied to the
links of the escapement, things not normally considered in our study
of mechanisms. First, the fourth wheel is not driven by the escape
wheel in Figure 8–7. Power is supplied by the mainspring through
the other wheels; the third wheel actually transmits this to the
fourth wheel. Thus, the escape wheel then only releases the fourth
wheel so that the torque applied by the third wheel can rotate the
fourth wheel.

Now consider the balance wheel and its motion. The
balance wheel oscillates at the frequency required for the escape
wheel and the associated gearing (the wheels) to show the correct
time. This rate of oscillation is determined by the mass of the
balance wheel and the design of the hairspring. A small amount of
energy is put back into the hairspring from the mainspring; it is
transmitted by the escape wheel during part of its cycle. Without
this, the oscillation from the hairspring would decay and eventually
stop.

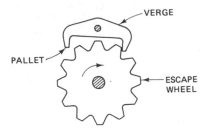

FIGURE 8–8. Verge, or runaway, escapement. Torque input, from spring or other source, is supplied to escape wheel, causing rotation. Verge oscillates as alternate pallets strike the escape wheel.

Clocks as well as watches use the balance wheel escapement. In the pendulum clock, however, the pendulum is the oscillating mass instead of the balance wheel. The escapement is similar to that used for the balance wheel.

Escapements are used in many other applications, and we shall discuss next some of the different escapement types. The *verge* escapement is shown in Figure 8–8. It is also called a *runaway* escapement. As explained in Figure 8–8, the torque input is directly to the escape wheel, resulting in continuous rotation of the escape wheel. As the teeth of the escape wheel strike the verge, the verge oscillates back and forth so that one pallet is always in contact with the escape wheel. The action of the verge escapement is different from that of the watch escapement; the verge tends to act as a speed control or damping device. It is used in timers, parking meters, bomb fuses, and other mechanisms.

Machine escapements are used to perform some of the same functions as Geneva mechanisms, ratchets, and cams. Figure 8–9 shows one design of a machine escapement. This escapement is spring-loaded so that the upper pallet is pulled down against the

FIGURE 8–9. Machine escapement.

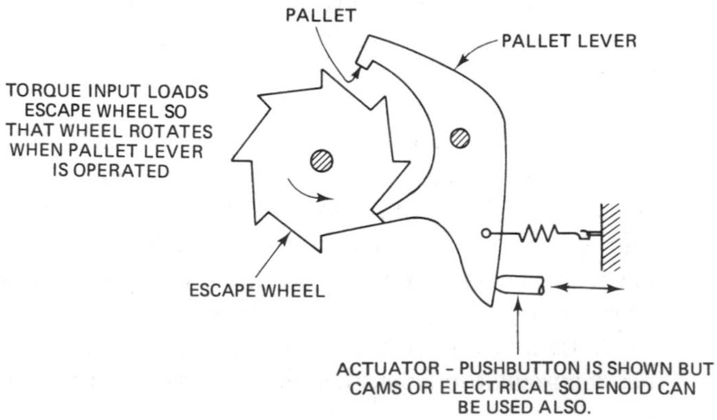

PALLET

PALLET LEVER

TORQUE INPUT LOADS ESCAPE WHEEL SO THAT WHEEL ROTATES WHEN PALLET LEVER IS OPERATED

ESCAPE WHEEL

ACTUATOR - PUSHBUTTON IS SHOWN BUT CAMS OR ELECTRICAL SOLENOID CAN BE USED ALSO.

escape wheel by the spring until it is moved by the actuator. The actuator then rotates the pallet lever to the position shown in Figure 8–9. With this type of escapement, the motion of the escape wheel is similar to that of the Geneva mechanism; it indexes to the next position, as does the Geneva. The difference between the two is that the input torque to the escapement is to the escape wheel, whereas in the Geneva mechanism the pin member is the driver and applies torque to the wheel.

8–5 STAR WHEELS

The *star wheel* is a mechanism that has some of the features of the Geneva mechanism. The drive is by pins as in the Geneva, but several pins are used in the star wheel instead of the one pin of the Geneva. Figure 8–10 shows one type of external star wheel design. Here the star wheel dwells for 180°. As the drive wheel rotates, the dwell surface of the drive wheel moves out from the star wheel, leaving pin 1 free to move into place and pick up slot 1 in the star wheel. Rotation continues with pin 2 picking up slot 2, and with each subsequent pin picking up its appropriate slot.

There are many variations of star wheel mechanisms, and internal star wheel mechanisms exist as well as external. A comparison of the star wheel in Figure 8–10 with the Geneva mechanism shows that the star wheel has one dwell for every revolution of the star wheel, or output, whereas the Geneva mechanism has a minimum of three dwells for every revolution of the output.

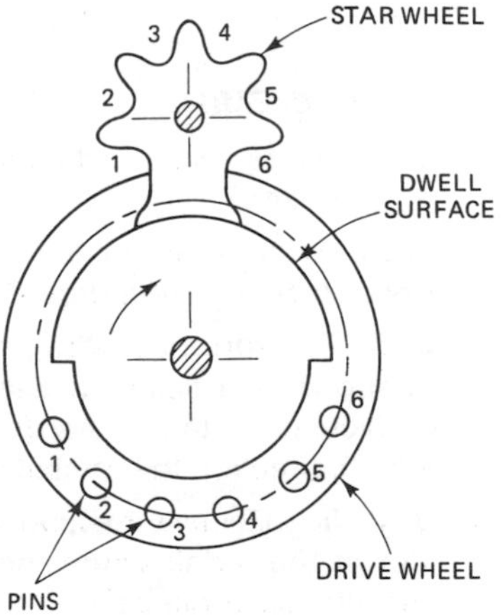

FIGURE 8–10. Star wheel mechanism.

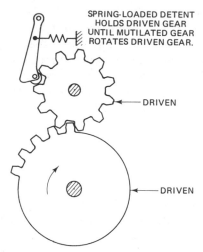

FIGURE 8–11. Mutilated gear mechanism.

8–6 MUTILATED GEARS

Still another type of intermittent motion mechanism is the *mutilated gear*. A mutilated gear is simply a gear with some of the gear teeth removed. Figure 8–11 shows a mechanism using a mutilated gear to obtain intermittent motion of the driven gear. An examination of Figure 8–11 will show that the number of dwells of the output is a function of the number of teeth on the driver as well as the diameters of the gears. Because of the impact that occurs when the driver first contacts the driven gear, mutilated gears are restricted to low speeds.

► PROBLEMS

8–1 Name four machines that use the ratchet as a part of their mechanism.

8–2 What is the advantage of the double-action ratchet over the other (single-action) types?

8–3 The horizontal straight link in Figure 8–12 moves back and forth as shown. It is desired to use this link to index the ratchet wheel above it. Sketch a suitable pawl design that would do this, and determine in which direction the ratchet wheel rotates.

8–4 In a Geneva mechanism the angle that the pin makes with the slot at the point where the pin enters the slot is important. What should this angle be?

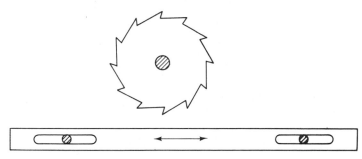

FIGURE 8–12

8–5 A Geneva mechanism is to have six stations. The distance between centers of the driver and the slotted wheel is 4 in. What is the required distance from the pin to the center of the driver?

8–6 In the four-station internal Geneva mechanism shown in Figure 8–6, what percent of the time does the slotted wheel dwell for each revolution of the driver?

8–7 A fundamental difference between the escapement and other mechanisms such as the Geneva is in the way that energy is applied to obtain mechanism motion. Explain this difference.

8–8 Makeup energy of the watch escapement is supplied by the _____. (Select the correct answer.) (a) hairspring (b) mainspring (c) pallet (d) none of these

8–9 What would happen if the makeup energy were not supplied to the watch escapement?

8–10 Star wheel mechanisms have some features in common with Geneva mechanisms. Name one of them.

8–11 Mutilated gears have one major drawback that restricts their use. What is it?

CHAPTER 9
CAMS

9–1 INTRODUCTION

The *cam* is one of the most important mechanisms in use. It is an *intermittent motion* mechanism in that the *cam follower* moves intermittently in response to the cam itself. The cam follower is another link in the mechanism train that always exists when a cam is used. The parts of the cam mechanism are identified in Figure 9–1. It is apparent that the shape of the cam itself determines the motion of the follower. The shape of the cam can be designed in an almost infinite number of ways, and this ability to provide so many shapes is what makes the cam so valuable.

The cam mechanism of Figure 9–1 is a common design. The follower reciprocates with straight-line motion in the frame, and its motions are called *rise* and *fall*, as indicated in Figure 9–1. Other types of followers are used; for example, the follower could be an arm pivoted about some other point on the frame.

9–2 CAM TYPES

Classification of cams by types is difficult because of the different methods used. However, two broad *classes* apply to all cams: (1) *two dimensional* and (2) *three dimensional*. *Two-dimensional* cams provide motion in two directions and in one plane. *Three-dimensional* cams provide motion in more than one plane and have three-dimensional coordinates. Chapter 2 mentioned briefly three-dimensional cams, and Figure 2–2 illustrates a three-dimensional cam.

Two-dimensional cams may be classified by the shape of the cam itself, by the kind of motion that the cam gives to the follower, and by the way that the cam holds or constrains the follower. These cam types are explained in Figures 9–2 through 9–4.

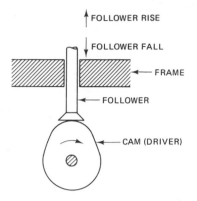

FIGURE 9–1. Common cam mechanism.

Four types of cams classified by shape are shown in Figure 9–2. The *plate* cam, also called the disk cam, is the most commonly used. The camshaft of the automobile engine is made up of a series of plate cams, with each cam through appropriate linkage opening and closing one valve in the engine. The automatic screw machine makes extensive use of plate cams as well as other types. When the centerline of the follower motion passes through the center of rotation of the cam, as it does in Figure 9–2(a), the plate cam is also called a *radial* cam.

FIGURE 9–2. Cam types classified by shape.

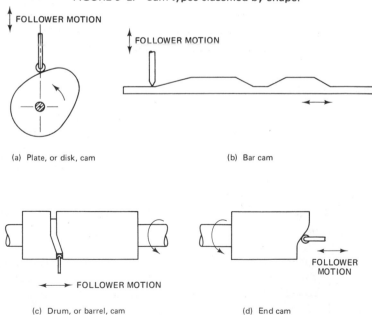

(a) Plate, or disk, cam

(b) Bar cam

(c) Drum, or barrel, cam

(d) End cam

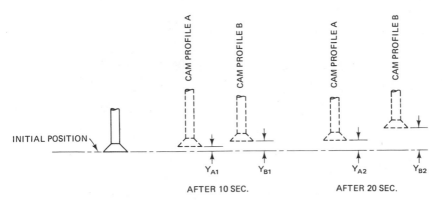

FIGURE 9–3. Hypothetical follower positions resulting from different cam profiles. Follower displacement, velocity, and acceleration characteristics may be used as a way to classify the type of cam.

The other types of cams in Figure 9–2, although not used as much as the plate cam, have many important applications from machine tools to typewriters.

Figure 9–3 shows hypothetical positions of a follower when it is moved by two cams with different profiles. Since the cam rotation is related to time, the movement of the follower is also time related. It therefore follows that displacement, velocity, and acceleration are follower characteristics, which in some way can be controlled by the cam profile design. *Constant velocity, constant acceleration, cycloidal*, and *simple harmonic motion* refer to the kinds of motion that the follower has. They are used also as a method of classifying cams. The type of cam motion is frequently specified along with the shape of the cam when describing the cam (for example, a plate cam with constant acceleration).

Examples of cams that restrain the follower motion are shown in Figure 9–4. The *face cam* is similar to the plate cam, but

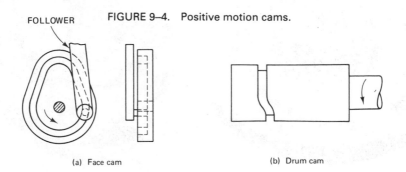

FIGURE 9–4. Positive motion cams.

(a) Face cam

(b) Drum cam

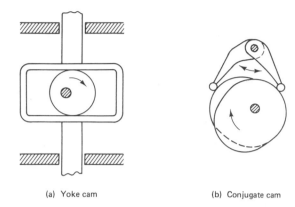

(a) Yoke cam   (b) Conjugate cam

FIGURE 9–5.   Other positive motion cams.

the follower motion is constrained to that of the groove. The *drum cam* also restrains follower motion. Both of these cam types are also called *positive motion cams*.

Figure 9–5 shows two additional types of positive motion cams. The method of restraint is different in these two cams. Instead of using a grooved track for the follower to ride in, as do the examples of Figure 9–4, the follower provides two surfaces to enclose the cam surface. As the cam rotates, motion is transferred first to one follower surface and then to the second. Note that the follower of the conjugate cam is a pivoted follower and has rotary motion instead of translation.

Photographs of commercial cams of some of these types are given in Figures 9–6 through 9–8.

FIGURE 9–6.   Split face cam. (Courtesy Ferguson Machine Company)

FIGURE 9–7.   Barrel cam. (Courtesy Ferguson Machine Company)

FIGURE 9–8. Conjugate cams. (Courtesy Ferguson Machine Company)

9–3 FOLLOWER TYPES

A primary consideration in follower design is the selection of the follower shape at the contact point on the cam. Examination of the followers in Figures 9–1 and 9–2 shows three types; they are the flat-faced follower in Figure 9–1, the roller follower in Figure 9–2(a), and the knife-edge follower in Figure 9–2(b). The roller gives the least *wear* of the three, and the knife-edge follower is the worst in this respect. *Wear* is an important consideration in design, and the roller follower is used extensively to reduce wear.

Because of the importance of the contact surface, followers are classified primarily by it. Second, the type description may include the location of the follower. Figure 9–9 explains the types of followers, starting with the roller follower. The roller follower of

FIGURE 9–9. Some follower types.

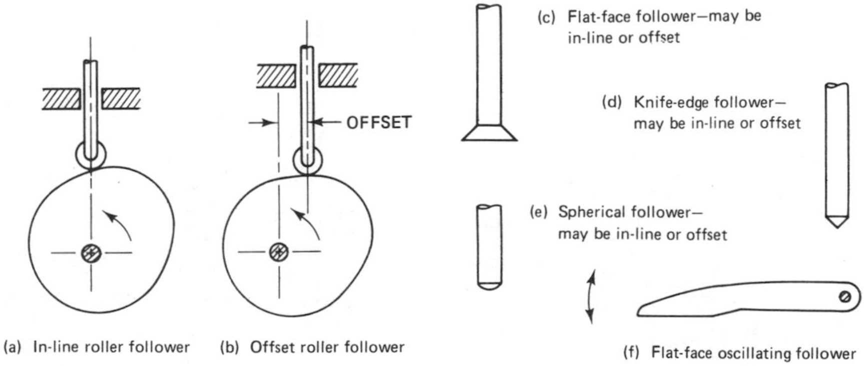

(c) Flat-face follower—may be in-line or offset

(d) Knife-edge follower— may be in-line or offset

(e) Spherical follower— may be in-line or offset

← OFFSET →

(a) In-line roller follower   (b) Offset roller follower

(f) Flat-face oscillating follower

Figure 9–9(a) is *in-line*, meaning that its centerline passes through the center of rotation of the cam. Figure 9–9(b) shows the *offset* roller follower; the centerline of the follower does not pass through the cam center of rotation. Generally, most followers are either in-line or offset. However, for any given motion, we shall see later that the cam profile is affected by changing to an offset follower.

9–4 CAM NOMENCLATURE

It is necessary to become familiar with cam nomenclature in order to proceed with the study of cam characteristics and design. We shall use the common plate cam of Figure 9–10 to illustrate the terms used in cam nomenclature. The follower for this cam is a roller follower.

The terms of Figure 9–10 are defined as follows:

Base circle. **The smallest-diameter circle that can be drawn from the center of rotation of the cam to the cam surface.**
Trace point. **The point that traces a curve that is defined by the displacement characteristics of the follower. For a knife-edge follower, the curve traced by the trace point is the profile of the cam.**
Pitch curve. **The curve traced by the trace point. For a knife-edge follower, it coincides with the cam profile.**
Prime circle. **The smallest-diameter circle that can be drawn from the center of rotation of the cam to the pitch curve.**

The *pressure angle* of a cam can be visualized by examining Figures 9–10 and 9–11. The cam in Figure 9–10 rotates counterclockwise and forces the follower up. The force of the cam on the

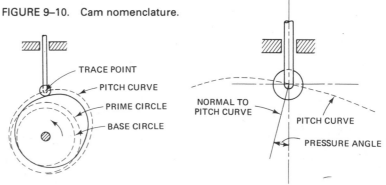

FIGURE 9–11. Pressure angle.

FIGURE 9–10. Cam nomenclature.

FIGURE 9–12. Undercut.

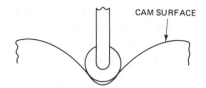

CAM SURFACE

follower is *normal* to the cam surface at the point of contact, but the follower is constrained so that it moves only in a vertical direction. The angle that the force makes with the follower direction of travel is called the *pressure angle*. Since the pitch curve has essentially the same form as the cam, the pressure angle is defined with respect to the pitch curve, as shown in Figure 9–11. It has the following definition:

The *pressure angle* is the angle between a normal to the pitch curve and the follower travel. The lines representing these directions are both drawn through the trace point.

The trace point for the roller follower is the center of the roller, as shown in Figure 9–11.

The pressure angle is important for several reasons. An excessive pressure angle creates a force component perpendicular to the follower travel. If this force component is too large, the follower tends to cock and jam in its ways, causing excessive wear. The roller itself may stop its rotation and tend to jam. In practice, designers try to keep the maximum pressure angle under 30°.

Undercutting of a cam is shown in Figure 9–12. The profile of the cam is such that the follower as designed here cannot follow it. The roller bridges a portion of the cam profile. The example of undercut shown here uses a roller follower. Less obvious cases of undercut exist with other types of followers, such as the flat-face follower. In this case a desired follower position may not be obtained when the cam profile necessary to provide adjacent follower positions does not contact the desired follower position. Increasing the cam size eliminates the undercut.

9–5 MOTION CHARACTERISTICS

The cam is designed to provide a predetermined and definite type of motion to its follower. To undertake this design, a study of the relationships existing among the displacement, velocity, and acceleration of the follower is necessary.

Chapter 2 noted the mathematical relationships existing among displacement, *s*, velocity, *v*, and acceleration, *a*, using calculus terms to write the equations. They are repeated here, with *t* being used to represent time.

$$v = \frac{ds}{dt} \qquad (2\text{--}3)$$

$$a = \frac{dv}{dt} \qquad (2\text{--}4)$$

Equation 2–3 states that velocity is equal to the derivative of displacement with respect to time. Equation 2–4 states that acceleration is equal to the derivative of velocity with respect to time.

These equations are the principles upon which our analysis of follower motion depends. The analysis uses graphical methods for simplicity, and, when carried through to include the cam profile layout, uses *inversion* as described in Chapter 1. The graphical layouts shown in Figure 9–13 illustrate how these are used.

FIGURE 9–13. Displacement, velocity, and acceleration graphs for constant velocity.

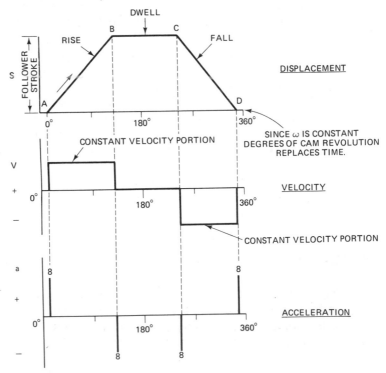

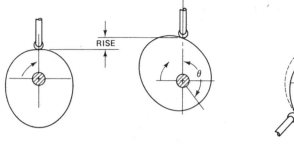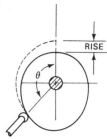

(a) Initial position of cam   (b) Cam rotated θ degrees   (c) Inversion follower rotated θ degrees instead of cam

FIGURE 9–14.   Inversion process.

In Figure 9–13 a *displacement diagram* is plotted for the follower. It represents the distance that the follower moves for a given angular rotation of the cam. *Since the angular velocity of the cam is constant, we can replace the time, t, of Equations 2–3 and 2–4 with the angle that the cam moves through.* The position of the follower is determined using *inversion*, as illustrated in Figure 9–14. Using this procedure in the graphical construction, the follower is rotated about the stationary cam as shown in Figure 9–14(c). Figure 9–14(a) and (b) shows the conventional rise of the follower as the cam rotates through an angle of θ degrees. The rise shown in Figure 9–14 thus corresponds to the vertical distance from the zero axis of the displacement diagram to the plot of the curve of Figure 9–13 at the angle θ.

It is fairly easy to visualize proper construction of the displacement diagram of Figure 9–13. The velocity and acceleration diagrams, however, have been included without any attempt to explain how they were constructed. We shall turn now to the reasoning that governed the construction of these.

When a *change* in displacement occurs, some finite value of velocity exists. The *change* in displacement is exemplified by the rise and fall portions of the displacement diagram of Figure 9–13. During the *dwell* portion of the diagram, the follower does not move; its velocity is therefore zero. To show the velocity existing at various cam positions, the velocity diagram is plotted in Figure 9–13 just under the displacement diagram. Use of the same horizontal scale for degrees of cam rotation as in the displacement diagram allows us to project down points where the follower motion and velocity change.

Now consider the rise portion of the displacement diagram. The curve is a straight line with a direction up and to the right.

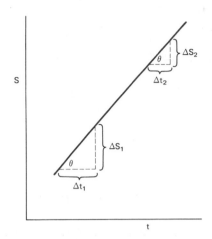

FIGURE 9–15. Constant rate of change. Rate of change of s with respect to t is represented by the slope of the curve. The slope is the tangent of the angle θ. For the increments of s and t shown, then $\tan \theta = \dfrac{\Delta s_1}{\Delta t_1}$. For any other increment Δs_2, $\tan \theta = \dfrac{\Delta s_2}{\Delta t_2}$. Since θ does not change, the rate of change of s is constant.

Because the curve is a straight line, the *rate of change* of the displacement is constant. Since the rate of change of displacement is also the velocity, the velocity during this portion is also constant. Figure 9–15 explains the statement that the rate of change of displacement is constant when the curve is a straight line.

In Figure 9–13, constant velocity is represented by the horizontal straight lines of the rise and fall portions of the velocity diagram. Note that the velocity during the fall of the follower is negative since its direction has been reversed.

The acceleration diagram can be explained by considering what happens at points A, B, C, and D on the displacement diagram, and also between these points. First, we have said that during the rise, fall, and dwell portions the velocity of the follower is either constant or zero. Thus, there is no change in velocity during these times. With no change in velocity, acceleration is zero. The acceleration diagram of Figure 9–13 thus is plotted with zero acceleration during these portions.

Now consider what happens at points A, B, C, and D. At all these points the velocity either goes *instantaneously* from zero to a finite velocity, or *instantaneously* from a finite velocity to zero. Theoretically, this means that the change in velocity is infinitely large. Thus, *acceleration* at these points is theoretically infinite. The vertical lines on the acceleration diagram represent the positive and negative infinite accelerations occurring at these points. Negative acceleration occurs when the velocity changes from either a positive or negative value to zero.

A cam having the characteristics of Figure 9–13 is called a *constant velocity cam*. The high accelerations at the points discussed result in high forces in the mechanism. For this reason the constant velocity cam is not used to a great extent.

The displacement, velocity, and acceleration diagrams are the basis for the cam design and layout discussed in Chapter 10.

► **PROBLEMS**

9–1 Explain why the cam is classified as an intermittent motion mechanism.

9–2 Which of the following cam types could be used satisfactorily with a flat-face follower? (a) end cam (b) drum cam (c) plate cam (d) bar cam

9–3 Define a positive motion cam and give two examples.

9–4 Which follower type would normally be selected for use with a cam when it is desired to keep wear of the cam and follower to a minimum? (a) flat-face (b) roller (c) knife-edge (d) spherical

9–5 Give one way to eliminate undercut occurring with a roller follower.

9–6 Some cams are classified by the type of motion that they give to the follower. Name two types of this motion.

9–7 Define pressure angle and tell why it is important in cam design.

9–8 What cam profile must always exist during the dwell portion of its rotation?

9–9 Figure 9–16 shows the displacement diagram of a cam. Identify the angular increments where the follower rise, dwell, and fall occur.

9–10 It is desirable to keep the pressure angle of a cam under _____ . (Select the correct answer.) (a) 35° (b) 30° (c) 40° (d) none of these

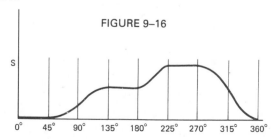

FIGURE 9–16

9–11 The trace point traces the pitch curve of the cam. In what type of follower does this pitch curve coincide with the cam profile?

9–12 The base circle of a cam is the smallest-diameter circle that can be drawn from the center of cam rotation to the _____. (Select the correct answer.) (a) pitch point (b) pitch curve (c) cam surface (d) none of these

CHAPTER 10
CAM DESIGN AND LAYOUT

10–1 INTRODUCTION

The design of a cam is governed by the type of follower motion required, the angular velocity of the cam, the precision required for the cam, the manufacturing method available to make the cam, and other factors dictated by the mechanism in which the cam is to be used.

Precision cam design and manufacture is a highly specialized field depending on sophisticated design techniques and specialized machine tools for manufacture. Design by computer and manufacture using numerically controlled machine tools are two techniques used for precision cams. *Tracer control* machining is also used; in this method the profile of a *master cam* is traced and reproduced on the part being made. With these methods it is possible to produce cams with profiles accurate to 0.0001 in.

The profile of a precision cam would ordinarily be determined analytically and then expressed in mathematical terms. However, for many cams, where precision is less important, the cam profile may be drawn by graphical means. The resulting layout then may be used as a pattern for making the cam. In this chapter we shall develop the graphical methods used for the layout and, to a lesser extent, also develop some analytical methods.

10–2 FOLLOWER MOTION

The follower motions used in this chapter are *constant velocity, modified constant velocity, constant acceleration, simple harmonic motion,* and *cycloidal motion.* Each motion has its own acceleration characteristics that indirectly govern the application of the cam. The definitions that follow describe these motions.

Constant velocity. The follower moves in such a way that it has a *constant* velocity at all times.

Modified constant velocity. The center portion of the follower motion (rise or fall) is constant velocity. Follower motion at the beginning or end of the rise or fall is *modified* to some other type of motion to decrease the large acceleration occurring at these points.

Constant acceleration. The follower moves in such a way that it has a constant acceleration at all times.

Simple harmonic motion. The follower moves so that it has *simple harmonic motion* at all times. *Simple harmonic motion* is motion of a point moving on a straight line in such a way that its acceleration is always proportional to the distance from the starting point and always directed toward the starting point.

Cycloidal motion. A cycloid is the path traced by a point on a circle as the circle rotates on a straight line, as shown in Figure 10–1. A follower having cycloidal motion follows the projection of this point, as shown in Figure 10–1.

These motions are described in more detail in later discussions relating to the displacement diagrams. Chapter 9 discussed in some detail the constant velocity cam, and the definition of simple harmonic motion was also given in Chapter 7.

10–3 DISPLACEMENT DIAGRAM CONSTRUCTION: CONSTANT VELOCITY

Although the displacement diagram for constant velocity follower motion was shown in Chapter 9, the procedure used in its construction is detailed here in the following steps.

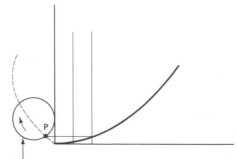

ROTATING CIRCLE

FIGURE 10–1. Cycloidal motion.

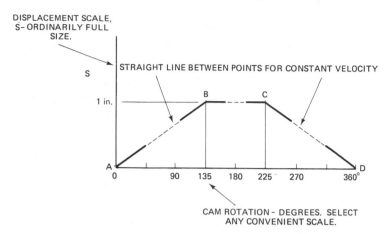

DISPLACEMENT SCALE, S- ORDINARILY FULL SIZE.

STRAIGHT LINE BETWEEN POINTS FOR CONSTANT VELOCITY

CAM ROTATION - DEGREES. SELECT ANY CONVENIENT SCALE.

FIGURE 10–2. Constant velocity displacement diagram.

Procedure

1. On the vertical scale, lay off the follower displacement to scale. The scale will ordinarily be full sized, since the displacement of the follower at various cam positions will be transferred from the displacement diagram to the cam layout. This is shown in Figure 10–2.
2. On the horizontal scale, lay off the degrees of cam rotation, from 0 to 360°. Any convenient scale may be selected.
3. From the prescribed follower motion, locate two points on the displacement diagram that represent the follower position at two different times. In Figure 10–2, two such points are *A* and *B*. Point *A* is at 0° and 0 rise; point *B* is at 135° and 1-in. rise. *Connect these two points with a straight line.* Points *C* and *D* are similarly connected for the fall portion of the follower motion. The *dwell* portion is a horizontal straight line. The complete follower motion as shown in Figure 10–2 is a rise of 1 in. from 0 to 135°, a dwell from 135 to 225°, and a fall from 225 to 360°.

10–4 DISPLACEMENT DIAGRAM CONSTRUCTION: CONSTANT ACCELERATION

Chapter 2 defined rectilinear motion of a point or body as motion in a straight line. Excluding the oscillating follower, the followers we have discussed have this type of motion. Chapter 4 also gave equations for constant acceleration of bodies having rectilinear motion. Equation 4–7, repeated here, relates the displacement, s, to acceleration.

$$s = v_o t + \tfrac{1}{2} a t^2 \qquad (4\text{–}7)$$

In this equation, the initial velocity is v_o, and t represents time.

Applying this to a follower, the initial velocity starting from rest is zero. Thus, the $v_o t$ term is zero. This leaves the following equation:

$$s = \tfrac{1}{2} a t^2 \qquad (10\text{–}1)$$

With constant angular velocity, the angle of cam rotation θ is proportional to the time, t. Therefore, θ^2 is also proportional to t^2. Since the acceleration is constant, we can make the following statement.

Displacement, s, is proportional to θ^2.

The construction of the displacement diagram makes use of this fact. The procedure for it follows.

Procedure

1. **Lay off the vertical and horizontal scales using the same methods as used for the constant velocity cam.**
2. **The cam angular displacement corresponding to the prescribed follower rise (or fall) is used to plot points on the curve. Divide this angular displacement interval into an even number of increments, as shown in Figure 10–3(a). The larger the number of increments, the more accurate the displacement curve and the cam will be.**
3. **Now locate the midpoint of the follower rise, as shown in Figure 10–3(a). Since each half of the curve is a mirror image of the other half, the midpoint of the displacement is used in plotting the curve.**
4. **In Figure 10–3, twelve increments, six for each curve half, have been used. These have been numbered 1 through 6 in Figure 10–3(b). The displacement, s, at each increment is proportional to the square of the angular displacement, θ. Now let the numbers 1 through 6 equal the values of θ at these points.**
5. **The vertical displacement corresponding to each numbered increment is proportional to the square of**

the number. For example, the displacement s_2 at increment 2 is proportional to 2^2, or 4. A graphical means to plot these points is shown in Figure 10-3(b).

6. Select a suitable scale with a number of divisions equal to the square of the number of increments corresponding to one half of the angular displacement interval, θ. For Figure 10-3(b), the number of increments is six and a scale with 36 divisions is used.

7. Lay off the 36 divisions on a line from the lower left corner of the graph, with the last division being on a horizontal line drawn through the rise midpoint. The angle that the 36-division line makes is not important.

FIGURE 10-3. Constant acceleration displacement program.

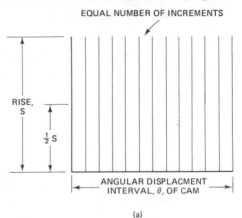

(a)

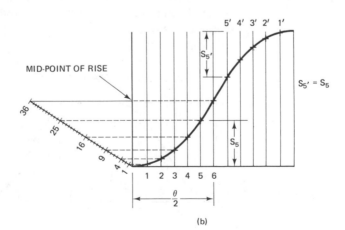

(b)

Number the points 1, 4, 9, 16, 25, and 36, as shown in Figure 10–3(b).

8. Project over horizontally these points until each one intersects its appropriate angular increment.
9. Draw a smooth curve through the points.
10. The other half of the curve may be drawn by transferring measurements, as shown in Figure 10–3(b). The angular increments are numbered in *reverse* order using primed numbers. The measurements transferred are measured from the top of the graph as shown; thus $s_5 = s_5'$.

The following example shows how this procedure is applied.

EXAMPLE

Plot the displacement diagram for a follower having constant acceleration and with the following motion.

$$0 \text{ to } 150°: \quad \text{rise } 1\tfrac{1}{2} \text{ in.}$$

$$150 \text{ to } 200°: \quad \text{dwell}$$

$$200 \text{ to } 360°: \quad \text{fall } 1\tfrac{1}{2} \text{ in.}$$

Solution

Figure 10–4 shows the solution for the problem. A suitable scale is selected to show the angular motion of the cam. It is divided into the rise, dwell, and fall portions as specified in the problem. In this case, the rise and fall portions are subdivided into 10 equal increments, leaving 5 increments for each half of the rise and fall portions. The vertical scale is full sized. A scale with 25 divisions, the square of 5, is selected, and the vertical displacement at each angular increment is obtained by the projection of points as done previously.

FIGURE 10–4. Displacement diagram.

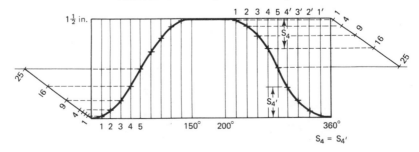

Note that the fall portion of the follower motion is constructed by using the upper right corner of the graph to locate the 25-division line.

10–5 DISPLACEMENT DIAGRAM: SIMPLE HARMONIC MOTION

Simple harmonic motion is motion of a point moving in a straight line in such a way that its acceleration is always proportional to the distance from the starting point and directed toward the starting point. A helical spring with a weight hanging on it gives simple harmonic motion when displaced, as shown in Figure 10–5. The motion is in a straight line, and the spring force on the weight always gives an acceleration toward the rest position of the weight, since it tends to slow down the weight as it goes past the rest position. Due to the spring force, the acceleration is also proportional to the distance.

This same type of motion can be duplicated by the projection of a point from a rotating line onto another line through the center of the rotation. This is shown in Figure 10–6. Here the line OP rotates about the center O. Point P moves in a circle. Its projection, P', onto a vertical line drawn through O gives simple harmonic motion.

FIGURE 10–5. Simple harmonic motion.

FIGURE 10–6. Simple harmonic motion. Line OP rotates about O. P' moves with simple harmonic motion at its various positions P', P'_1, and P'_2.

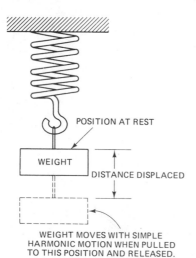

POSITION AT REST

WEIGHT

DISTANCE DISPLACED

WEIGHT MOVES WITH SIMPLE HARMONIC MOTION WHEN PULLED TO THIS POSITION AND RELEASED.

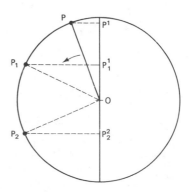

The motion of P' in Figure 10–6 is the basis for the construction used to plot the displacement diagram for simple harmonic motion. Its procedure follows.

Procedure

1. **Lay off the vertical and horizontal scales in the same manner as used previously.**
2. **Divide the cam angular displacement into the proper intervals representing the rise, dwell, and fall of the follower.**
3. **For the rise portion, divide the cam angular displacement interval into a number of equal increments. For ease of construction, this number is usually an even number. This step is shown in Figure 10–7(a).**
4. **At the left side of the graph, construct a semicircle with diameter equal to the follower rise. This is also shown in Figure 10–7(a). Divide this semicircle into the same number of equal increments that was used for the angular displacement. This step is also shown in Figure 10–7(a), and the increments are numbered in a clockwise direction starting from the bottom.**

FIGURE 10–7.   Simple harmonic motion construction.

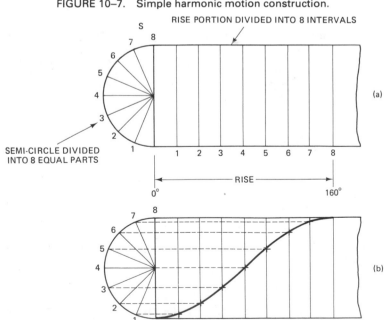

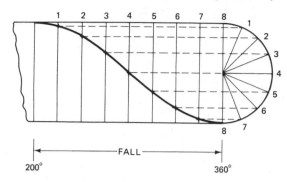

FIGURE 10–8.   Follower fall with simple harmonic motion.

5. **Project the points from the semicircle horizontally over to the appropriate vertical line representing the angular displacement. This is shown in Figure 10–7(b). Draw a smooth curve through the points.**
6. **To construct the fall portion of the follower motion, draw the semicircle on the right of the graph, as shown in Figure 10–8. The semicircle increments are numbered from the top and then projected over as shown.**

10–6 DISPLACEMENT DIAGRAM CONSTRUCTION: CYCLOIDAL

Figure 10–1 illustrates cycloidal motion. It is the path traced by a point on a circle as the circle rolls on a straight line. This type of motion can be applied to follower motion if we make the circumference of the rolling circle equal to the follower displacement, as illustrated in Figure 10–9. Here the circumference of the rotating circle has been made equal to the follower rise, s. The point P must travel a vertical displacement equal to s as the circle makes one revolution. The projections of the point P onto the vertical lines, as shown in Figure 10–9, determine the follower motion.

The procedure for the construction of the displacement diagram uses the circle with circumference equal to displacement in a somewhat different manner from that shown in Figure 10–9. The procedure follows.

Procedure

1. **To construct the circle, the radius, r, is used. Remembering that the circumference of a circle is 2π multiplied**

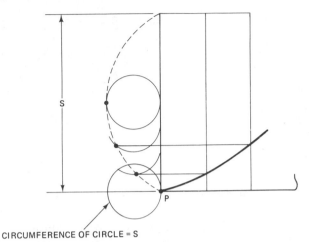

CIRCUMFERENCE OF CIRCLE = S

FIGURE 10–9.   Cycloidal motion of point *P*.

by the radius, and that the circumference is equal to the rise, *s*, then

$$s = 2\pi r$$

and

$$r = \frac{s}{2\pi} \tag{10–2}$$

2. Lay out the vertical and horizontal scales.
3. Divide the cam displacement interval into an even number of increments, as shown in Figure 10–10(a). Number the increments as shown.
4. Now lay off the diagonal line *AB*, as shown in Figure 10–10(a). At a convenient location near the lower left corner of the graph, draw the circle with radius *r*. In Figure 10–10, a rise of 3 in. has been used. The calculated value of *r* is 0.477 in., as shown in Figure 10–10(a).
5. Divide the circle into the same number of equal parts as used for the cam displacement interval. This is shown in Figure 10–10(b); note that the divisions start at the point 0 and go clockwise around the circle. Number the divisions as shown.
6. Project horizontally onto a vertical line through the circle center each numbered point on the circle circumference. This is shown in Figure 10–10(b). Points 0, 4, and 8 on the circumference when so projected converge at the center of the circle.

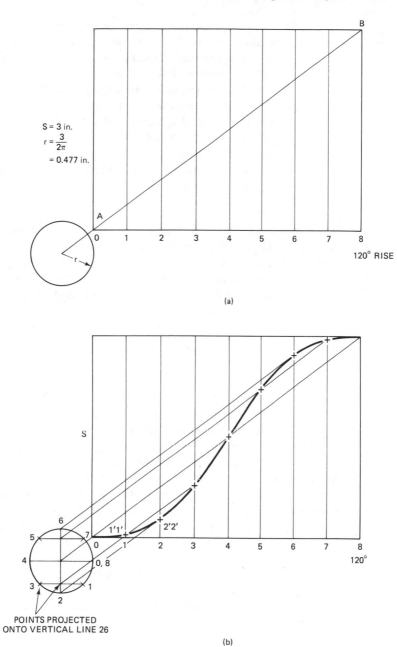

(a)

(b)

FIGURE 10–10.   Cycloidal follower motion.

7. **The points on the displacement curve are the intersections of the diagonal and the appropriate vertical lines and the intersections of parallels to the diagonal and the vertical lines. This construction is shown in Figure 10–10(b). For example, point 1'1' is the intersection of a line parallel to the diagonal and the vertical line at the cam displacement increment 1. The parallel lines pass through the appropriate projection on the vertical line through the circle center.**

8. **When all points are obtained, draw a smooth circle for the displacement curve.**

Full-scale graphical layouts for cycloidal motion are limited in use because of inaccuracies resulting when the follower rise is small. The example illustrated has a 3-in. rise and a circle radius of 0.477 in. A 1-in. rise would give a circle radius of one third of this, or 0.159 in. This radius is too small for accurate full-sized construction. For this reason, we shall develop analytical methods for cycloidal motion later in this chapter.

10–7 DISPLACEMENT DIAGRAM: MODIFIED CONSTANT VELOCITY

The limitations of the constant velocity cam were discussed in Chapter 9, and the high acceleration points shown in Figure 9–15. As the name implies, the *modified* constant velocity cam is modified to reduce these high accelerations to acceptable values. This modification consists of combining part of the constant velocity displacement curve with that of another type of motion having more acceptable characteristics. The effect on the displacement diagram is shown in Figure 10–11. The start and stop of follower motion, where high acceleration occurs, has been modified by introducing a different curve at these points.

Constant acceleration motion is frequently used to modify the constant velocity curve. Its characteristics allow the curve to be tangent to the straight line at the junction point. The construction of the modified constant velocity follower displacement diagram is a combination of that used for both motions. Its procedure follows.

Procedure

1. **Lay out the vertical and horizontal scales, and locate the cam displacement interval for the desired follower rise.**

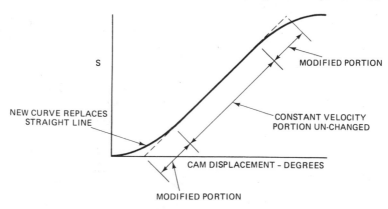

FIGURE 10–11.   Modified constant velocity motion.

2. **Divide the cam displacement interval into three parts: constant acceleration, constant velocity, and constant acceleration. This is shown in Figure 10–12(a). The proportions allotted to the various types of motion may vary, but obviously there has to be sufficient displacement allowed for the constant acceleration portion in order for it to be effective. In Figure 10–12(a), the follower rise occurs over 140° of cam displacement. The first 30° and the last 30° of cam displacement have been chosen for the constant acceleration portions of the rise.**

3. **Locate points *A* and *B* at the *midpoints* of the constant acceleration portions, as shown in Figure 10–12(a). Then connect the two points by a straight line to provide the constant velocity curve. Location of points *A* and *B* at the midpoints provides a location of the constant *velocity* curve that ensures that the constant *acceleration* curves are tangent when constructed. (This can be proved mathematically, but its proof is left to more advanced texts.)**

4. **Construct the constant acceleration portions using the methods previously outlined. This is done in Figure 10–12(b).**

10–8   DISPLACEMENT DIAGRAMS BY ANALYTICAL METHODS

Analytical methods for determining follower displacement are used when graphical methods do not provide sufficient accuracy. We

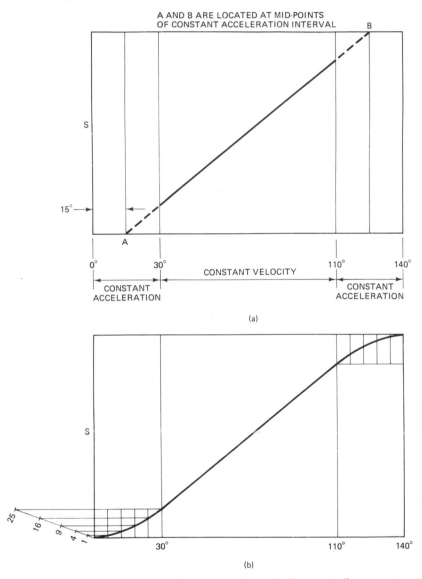

FIGURE 10–12. Modified constant velocity displacement diagram.

shall cover here the analytical methods used for two types of motions, simple harmonic and cycloidal.

Follower displacement can be calculated if we know the equation for the type of motion used. The equations use the terms shown in Figure 10–13. The coordinates of any point P are y, the vertical displacement, and angle θ, where θ is the angle that the cam has rotated through. The angle β (beta) is the cam displacement interval corresponding to the rise.

The derivation of the equations for simple harmonic motion and cycloidal motion is reasonably complicated and is not given here. The equations for each of these motions follow.

Simple Harmonic Motion

$$y = \frac{s}{2}\left(1 - \cos\frac{\pi\theta}{\beta}\right) \tag{10-3}$$

Cycloidal Motion

$$y = \frac{s}{\pi}\left(\frac{\pi\theta}{\beta} - \frac{1}{2}\sin\frac{2\pi\theta}{\beta}\right) \tag{10-4}$$

The angles $\pi\theta/\beta$ and $2\pi\theta/\beta$ in these equations are in radians. However, θ and β may be substituted in the terms using degrees, since the term θ/β is a ratio and all units cancel.

To use these equations for the *fall* portion of the follower motion, simply reverse the usual left-to-right numbering system and proceed as shown in Figure 10–14. The equations may be used in different ways to obtain the vertical displacement, *y*. Some texts use partially precalculated tables in which some of the terms of the equations have been calculated based on different ratios of the cam angle θ to the cam displacement interval β. The most straightforward method is to substitute different values of θ into the equation and solve for *y*. This method is used in this text.

FIGURE 10–13. Terms for motion equations.

FIGURE 10–14. Use of terms for fall portion.

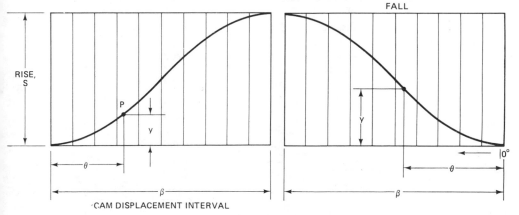

EXAMPLE

A cam is to be designed for cycloidal motion. Its follower is to rise 4 cm from 0 to 120°, then dwell from 120 to 220°, then fall from 220 to 330°. It dwells from 330 to 360°.

Plot the displacement diagram using calculated values for the displacement.

Solution

Since the follower displacement is to be calculated, it is assumed that an accurate cam profile is desired. Accuracy is improved by selecting small increments for the cam angle θ. Ten divisions of the rise will give 12° increments and will be used. Since the calculations are rather lengthy, only sample calculations are shown here. The results are included in a table for ease in plotting.

For the rise portion, the calculations and table follow. The equation is

$$y = \frac{s}{\pi}\left(\frac{\pi\theta}{\beta} - \frac{1}{2}\sin\frac{2\pi\theta}{b}\right) \qquad (10\text{–}4)$$

where $\beta = 120°$, $s = 4$ cm, and θ increases in 12° increments. The equation can be simplified if s is substituted into it as follows:

$$y = \frac{4}{\pi}\left(\frac{\pi\theta}{\beta} - \frac{1}{2}\sin\frac{2\pi\theta}{\beta}\right)$$

$$= 4\frac{\theta}{\beta} - 0.6366\sin\frac{2\pi\theta}{\beta}$$

The table is set up using all terms of the equation.

θ (degrees)	$4\dfrac{\theta}{\beta}$	$0.6364\sin\dfrac{2\pi\theta}{\beta}$	y (cm)
12	0.4	0.3742	0.026
24	0.8	0.6054	0.195
36	1.2	0.6054	0.595
48	1.6	0.3742	1.226
60	2.0	0.0000	2.000
72	2.4	−0.3742	2.774
84	2.8	−0.6054	3.751
96	3.2	−0.6054	3.805
108	3.6	−0.3742	3.974
120	4.0	0.0000	4.000

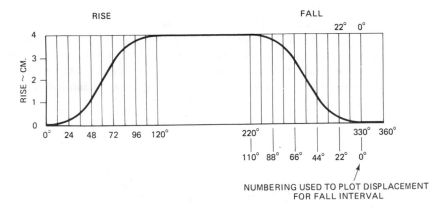

FIGURE 10–15

The value of each term in the table has been calculated for each of the cam angles, θ. The displacement y is then obtained by algebraically subtracting terms.

Note that the term $2\pi\theta/\beta$ is always in radians, and that the trigonometric sign convention applies; that is, the sine of an angle in the third and fourth quadrants is negative. This explains the negative values of the table.

The preceding table gives the displacement values for the *rise* of the follower. Now consider the *fall* portion of the follower motion. The follower falls from 220 to 330°, a 110° interval. If we use the same number of increments, 10, as used for the rise, the values of y are the same as the rise. (NOTE: The reader may wish to check this statement by performing sample calculations.)

It now remains to plot the displacement curve. This is done in Figure 10–15. Note that the plot of the fall portion uses the reversed angle numbering, as mentioned previously.

10–9 COMPARISON OF CAM CHARACTERISTICS

The cams discussed to this point have varying characteristics that affect their end use. In particular, the acceleration characteristics of a cam are important in cam selection. In this section we shall compare the acceleration and velocity characteristics of the cam previously discussed.

Figures 10–16 through 10–20 show representative graphs of displacement, velocity, and acceleration for the five types of motion that we have discussed. The graphs are not to scale and are intended to show only the *shape* of the curves. Except in the case of infinity, no graph should be used for judging the magnitude of any quantity.

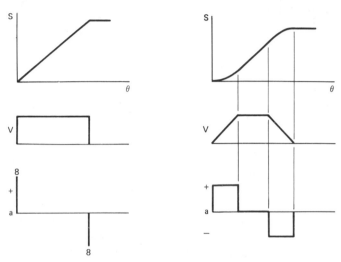

FIGURE 10–16. Constant velocity motion.

FIGURE 10–17. Modified constant velocity motion.

FIGURE 10–18. Constant acceleration motion.

FIGURE 10–19. Simple harmonic motion.

FIGURE 10–20. Cycloidal motion.

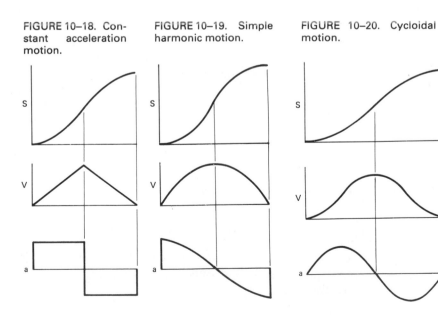

Figure 10–16 shows again the displacement, velocity, and acceleration curves for the *constant velocity cam*. The disadvantage is the high acceleration at the ends. The acceleration theoretically goes to infinity, resulting in destructive loads on the mechanism.

The constant velocity cam can be made into a practical device by modifying the motion at the ends. The characteristics of the *modified constant velocity* cam are shown in Figure 10–17. The end motion has been modified to constant acceleration motion. The acceleration values are constant and finite, as shown in Figure 10–17. However, because of the sharp change in *acceleration*, the modified constant velocity cam is restricted to low speeds. The sudden change in acceleration produces shock in the machine.

The *constant acceleration motion* shown in Figure 10–18 is also the motion used with the modified constant velocity cam, and the same comments about the sudden change in acceleration also apply.

Simple harmonic motion, shown in Figure 10–19, has a sudden change in acceleration at two points, the beginning and end of motion. This compares with the three points (beginning, midpoint, and end) of constant acceleration. However, this change in acceleration still restricts simple harmonic motion to low-to-medium speeds.

There are no sudden changes in acceleration in the curves for *cycloidal motion*, shown in Figure 10–20. It is thus suitable for high speed and provides the smoothest motion of all the types discussed.

10–10 CAM LAYOUT: GENERAL

The base circle diameter and the displacement diagram essentially provide the cam design. If these two are given, we now can proceed with the cam *layout*.

The cam layout is made full sized, except in some special cases when the layout has to be enlarged or reduced in size by photographic or other methods. It is desirable to keep the cam size small in most cases, but pressure angle and undercutting considerations may dictate changes in the size.

In making the layout, *inversion* is used. The follower is made to rotate about the cam by drawing it in different positions around the cam. The *trace point* traces the *pitch curve* using displacement values from the displacement diagram of the follower. Figure 10–21 illustrates one method of transferring displacement

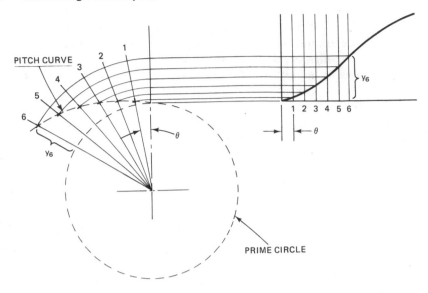

FIGURE 10–21. Transfering follower displacement to the prime circle.

values directly to the cam prime circle. Although the illustration is self-explanatory, we shall list some of the more important steps in the construction.

1. The prime circle must be positioned so that its diameter lies on a line extended from the base of the displacement diagram.
2. Points on the displacement curve are projected horizontally over to intersect with a vertical line through the cam center.
3. Starting in a direction *opposite* to the cam rotation, radial lines are drawn out from the cam center. *The angle at which the radial line is drawn is the cam angle θ used in the displacement diagram.*
4. A point on the pitch curve is obtained by striking an arc from the displacement on the vertical line to the appropriate radial line. Numbering the increments as shown is an aid in ensuring a correct plot.

This technique is used successfully when a graphical displacement plot is used. If the displacement values have been calculated analytically, each displacement is measured directly on the appropriate radial line starting from the prime circle.

Except in the case of the knife-edge follower, the pitch curve is not the cam profile. The cam profile is obtained by inversion, using the follower and rotating it around the cam. This

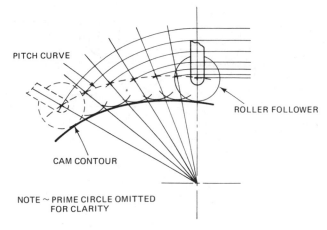

PITCH CURVE

ROLLER FOLLOWER

CAM CONTOUR

NOTE ~ PRIME CIRCLE OMITTED
FOR CLARITY

FIGURE 10–22.   Plotting cam contour by inversion.

principle is illustrated in Figure 10–22, which shows the same portion of pitch curve illustrated in Figure 10–21, but with some construction details removed. The roller follower used here is drawn in its initial position and then rotated around the cam, in this case counterclockwise. The center of the roller is located on the radial line at each position, and an arc is drawn to show the roller. The cam contour is obtained by drawing a smooth curve tangent to the roller arcs.

Although inversion of the follower is used to obtain the cam profile in all cases, the details of construction vary with the type of follower used and also with whether it is an in-line or offset follower. In the sections following, we shall show the construction required for the more commonly used follower types.

10–11   LAYOUT FOR THE KNIFE-EDGE, IN-LINE FOLLOWER

This is the simplest layout to make because the pitch curve is the cam profile. We shall show its construction by means of the following example.

EXAMPLE

Lay out the plate cam for the follower motion shown in the displacement diagram in Figure 10–23. The follower is a knife-edge, in-line follower, and is to rise 0.85 in. from 0 to 160°, dwell from 160 to 240°, and fall 0.85 in. from 240 to 360°. The base circle diameter is 3 in., and the cam rotates clockwise.

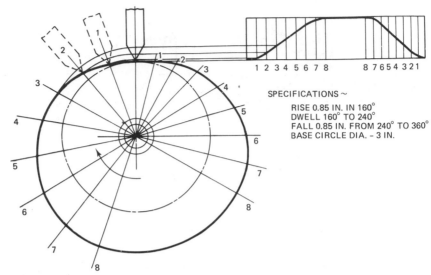

SPECIFICATIONS ~

RISE 0.85 IN. IN 160°
DWELL 160° TO 240°
FALL 0.85 IN. FROM 240° TO 360°
BASE CIRCLE DIA. – 3 IN.

FIGURE 10–23.   Layout for knife-edge, in-line follower.

Solution

The layout is shown in Figure 10–23. The follower is shown rotated *counter*clockwise (we noted earlier that the follower is rotated opposite to the cam rotation) into three different positions. Since the pitch curve and cam contour coincide for the knife-edge in-line follower, it is not necessary to redraw the follower at each position.

In Figure 10–23, only the first three points are shown projected over. Construction for remaining points is omitted for clarity.

10–12   LAYOUT FOR THE IN-LINE ROLLER FOLLOWER

The in-line roller follower layout is used for the illustration in Figure 10–22. It requires drawing the roller profile in different positions and matching the cam contour to the roller. Its construction is illustrated in the following example.

EXAMPLE

Construct the cam profile for the same displacement diagram used for the in-line knife-edge follower. The follower motion is the same, and the cam also rotates clockwise. The roller is $\frac{1}{2}$ in. in diameter.

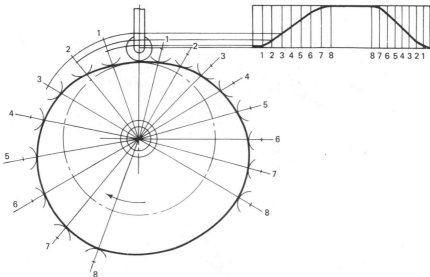

FIGURE 10–24.   Layout for in-line roller follower.

Solution

Figure 10–24 shows the construction. The points on the pitch curve are obtained in the same manner as previously done. At each point an arc equal to the radius of the roller is swung. The cam profile is then constructed with a smooth curve tangent to the arcs.

10–13 LAYOUT FOR THE OFFSET ROLLER FOLLOWER

The construction for the offset roller follower becomes more complicated because the roller has to be drawn in its offset position at each angular cam position. Again we shall use an example to illustrate how this is done.

EXAMPLE

Using the same specifications and displacement diagram used in the last two examples, layout the plate cam for an offset roller. The roller is offset $\frac{1}{2}$ in. to the left, and the roller diameter is $\frac{1}{2}$ in.

Solution

Figure 10–25 shows the cam layout. The bottom of the displacement diagram and the roller center are on the same

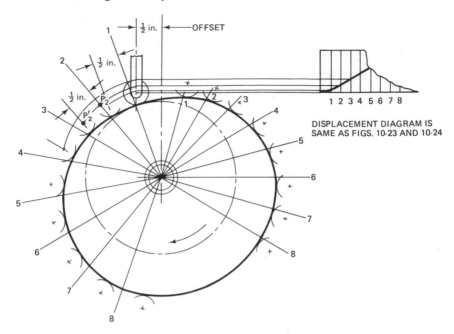

FIGURE 10–25. Layout for offset roller follower.

horizontal line when the follower is at its lowest position. After positioning the roller and displacement diagram to obtain this, the vertical centerline of the cam is located $\frac{1}{2}$ in. to the right of the roller centerline. A base circle diameter of 3 in. has been used on the other cams and will be used here. Using a radius of $1\frac{1}{2}$ in., draw a circle with its center on the vertical centerline of the cam and tangent to the roller. This step locates the horizontal centerline of the cam.

Follower displacements are projected over to the *follower* centerline. The points of intersection are then rotated around to the appropriate cam radial line. *The intersection point on the radial then must be transferred counterclockwise $\frac{1}{2}$ in. to provide the offset.* The $\frac{1}{2}$-in. dimension is on a line perpendicular to the radial line; one method of determining the final location of the trace point is to use a parallel line located $\frac{1}{2}$ in. from the radial. This is illustrated in Figure 10–25 by point P_2. The final trace point P_2' is the intersection of the parallel line and the arc from the cam center through P_2. Circular arcs are then drawn at each trace point, and the cam contour is constructed tangent to the arcs.

10–14 LAYOUT FOR THE FLAT-FACED FOLLOWER

The layout of a cam using a flat-faced, in-line follower is shown in Figure 10–26. The procedure used is similar to that for the roller follower, where the cam contour is drawn so that it is tangent to the roller at its various positions. In this case, the follower profile is a flat surface perpendicular to the radial line from the cam center. In examining Figure 10–26 it will be noted that contact of the cam with the follower occurs at the center of the follower only at 0° of cam rotation and at the dwell portions. This means that the flat follower surface has to be large enough to bridge the cam contour.

The cam in Figure 10–26 has different specifications from those used previously. The rotation is counterclockwise, and the rise and fall portions are different, as is the total amount of rise.

Construction of the displacement diagram and location of the follower at its lowest point are done with the usual procedure. Since the cam rotation is counterclockwise, the follower inversion is in the clockwise direction. Points on the displacement diagram are projected over to the vertical line through the follower center, and then rotated clockwise to the appropriate radial. At the point where

FIGURE 10–26. Layout for a flat-faced in-line follower.

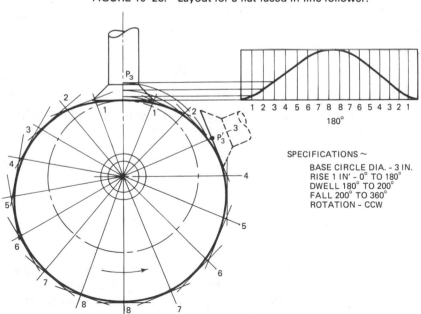

1 2 3 4 5 6 7 8 8 8 7 6 5 4 3 2 1
180°

SPECIFICATIONS ~

BASE CIRCLE DIA. - 3 IN.
RISE 1 IN' - 0° TO 180°
DWELL 180° TO 200°
FALL 200° TO 360°
ROTATION - CCW

the radial is intersected, a perpendicular to the radial is erected. The construction showing point P_3 rotated to P_3' illustrates this. At this point the position of the follower is shown in dashed lines. In practice, it is only necessary to draw the perpendicular at each position. The cam contour is drawn in tangent to all the perpendiculars as the last step.

10–15 LAYOUT FOR AN OSCILLATING ROLLER FOLLOWER

All the cams that we have drawn to now have had followers with rectilinear translation motion. The layout for an oscillating follower is similar to that for the translating follower, except that the desired follower motion occurs in an *arc* instead of on a straight line. This requires some changes in the construction methods we have used previously.

Figure 10–27 shows a cam with an oscillating roller follower. In examining this, it is obvious that the arc length, s, is affected by the angle ϕ and the arm length, L. Mathematically, the length of any circular arc is equal to the radius of the arc multiplied by the central angle in radians. Using the terms of Figure 10–27, we can write an equation for the arc length, as follows:

$$s = L\phi \qquad (10–5)$$

where

$$L = \text{arm length}$$

$$\phi = \text{central angle in radians}$$

Equation 10–5 can be changed to use degrees instead of radians in the angle. It becomes

$$s = \frac{\pi\phi L}{180} \qquad (10–6)$$

where

$$\phi = \text{central angle in degrees}$$

$$L = \text{arm length}$$

Referring to Figure 10–27, the rise increments from the displacement diagram are measured as chordal lengths, as indicated by the notation in Figure 10–27. This makes it impossible to project displacements from the displacement diagram, and the measure-

FIGURE 10–27. Cam with
oscillating roller follower.

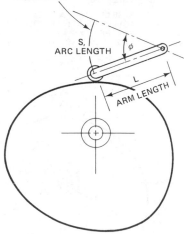

RISE INCREMENTS ARE MEASURED AS
CHORDAL LENGTHS ON THE ARC, USING
SHORT INCREMENTS FOR INCREASED
ACCURACY.

S,
ARC LENGTH

φ

L

ARM LENGTH

ments are transferred by dividers. The following example illustrates
the construction of a cam with an oscillating roller follower.

EXAMPLE

Lay out a plate cam with an oscillating roller follower. The
cam rotation is counterclockwise, the pitch circle diameter
is 3 in., the rise is 1 in., and the arm length is 3 in. The
displacement of the follower is shown in the displacement
diagram of Figure 10–28. Diameter of the roller is $\frac{1}{2}$ in.

Solution

The construction of the cam is also shown in Figure 10–28.
Note that the location of the displacement diagram on the
drawing is not important, since displacements are not
projected over to the cam as done previously.

To proceed with the construction, the base circle is
first drawn with the roller in its lowest position. The displacement diagram shows six cam angular rotation intervals for the 180° rise and also six for the 180° fall, giving 30°
for each interval. The base circle is then divided into these
30° intervals. Since the cam rotates counterclockwise, the
roller inversion is clockwise, and the radial lines forming
the intervals are numbered as shown.

Next construct the follower arm in place and the
arc representing the rise. The arc length *s* has been given

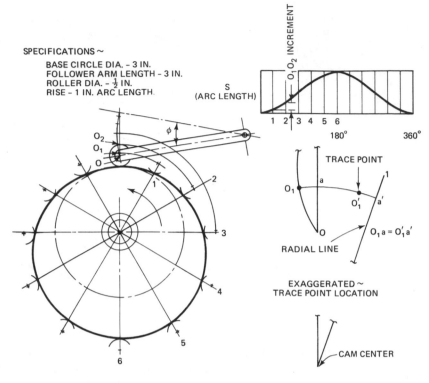

SPECIFICATIONS ~
    BASE CIRCLE DIA. – 3 IN.
    FOLLOWER ARM LENGTH – 3 IN.
    ROLLER DIA. – ½ IN.
    RISE – 1 IN. ARC LENGTH.

FIGURE 10–28.   Layout for an oscillating roller follower.

and the arm length also; it now remains to find the angle ϕ. This is done using Equation 10–6,

$$s = \frac{\pi \phi L}{180}$$

Rearranging and solving for ϕ,

$$\phi = \frac{180s}{\pi L}$$

Substituting,

$$\phi = \frac{180(1)}{\pi(3)}$$

$$= 19.1°$$

The center of follower rotation can be located by drawing a line through the roller center at an angle $\phi/2$ with the horizontal and measuring out 3 in. on the line. The centerline of the follower in its uppermost position is then drawn in, and the arc of the roller center travel is struck. *It is logical to locate the center of rotation of the follower arm to the right of the cam when counterclockwise rotation is used.* With the arm in this trailing position, forces on the roller acting through the pressure angle are smoother and cannot cause jamming.

Now the increments of displacement are transferred to the arc. The chordal length oo_1 is the displacement increment at cam interval 1. The increment at cam interval 2 is measured from point o_1 to point o_2 on the arc, and the next point obtained in the same manner. It is necessary to use incremental displacements instead of the total displacement to provide sufficient accuracy, since there is little error when small chordal lengths are used to approximate arc lengths.

The *trace point* of the cam is the roller center. Except in the lowest and highest follower positions, it is located *off* the radial line corresponding to the cam interval, as shown in the exaggerated view of Figure 10–28. After all trace points have been located and the roller radii drawn, the cam contour is drawn in.

10–16 DISCUSSION

We have covered some of the basic cam types and the methods used to construct the cams when the design specifications are given. Other points about cams are worth discussing; our discussion of these will be general and may or may not apply to all cases.

In the positive motion cam, the follower is constrained in both directions of movement by the cam itself. In all the other cam types, we have shown that the follower is in contact with the cam surface, and the assumption has been made that it remains this way. In practically all cases, however, the mechanism train has to be designed with a member to do this. Usually this member is a spring, and the valve spring of the automobile engine is a good example. The valve exerts a force through the rocker arm linkage that holds the follower in contact with the cam. Figure 10–29 shows another way in which a spring might be used to maintain follower and cam contact.

Ordinarily, it is desirable to design the cam with a small base circle diameter, but sometimes this may give an excessive

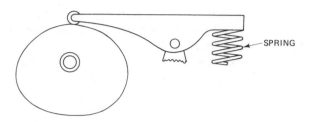

FIGURE 10-29

pressure angle. If this occurs, it is necessary to redesign to eliminate the condition. Increasing the cam size will help, and change may also be made in the follower motion and offset (if used) to improve the condition. One advantage of the flat-faced follower is that it is not critical to high pressue angles.

Increasing the cam size is also a solution to undercutting. If a roller follower is used, decreasing the roller size will also help in eliminating the undercutting.

► PROBLEMS

10–1 Select the type of follower motion that best fits the applications described in the following.

(a) The cam rotates at 880 rpm and the follower must have constant velocity at the midpoint of its travel.

(b) The cam rotates at 4000 rpm; the follower has a 5-cm rise and must have minimum shock.

(c) The cam rotates at 300 rpm, and the follower at the midpoint of its travel has zero acceleration. The acceleration at 1 cm on each side of the midpoint is 2 m/sec^2, and 2 cm each side of the midpoint it is 4 m/sec^2.

10–2 Construct the displacement diagram for a follower with the following motion: rise 3.5 cm from 0 to 180° with constant velocity; fall 3.5 cm from 180 to 360° with constant velocity.

10–3 Lay out the cam for the displacement diagram of Problem 10–2 using the following specifications: the follower is knife-edge, in-line, rotation is clockwise, and the base circle diameter is 8 cm.

10–4 Lay out the cam in Problem 10–3, except use an in-line roller follower of 1-cm diameter.

10–5 Construct the displacement diagram for a follower that rises $1\frac{1}{2}$ in. from 0 to 160° with modified constant velocity, dwells from 160 to 180°, and falls from 180 to 360° with modified constant velocity. Make the center one third of the follower travel constant velocity and modify the end thirds.

10–6 Lay out the cam for the displacement diagram of Problem 10–5 using the following specifications: base circle diameter is 4 in., the rotation is counterclockwise, and the follower is an in-line roller follower with a $\frac{1}{2}$-in.-diameter roller.

10–7 Lay out the cam in Problem 10–6, except offset the roller $\frac{1}{2}$ in. to the right.

10–8 Lay out the cam of Problem 10–6 using a flat-faced in-line follower.

10–9 Construct the displacement diagram of a follower having the following motion: rise $1\frac{1}{4}$ in. from 0 to 150° with constant acceleration, dwell 150 to 200°, fall from 200 to 360° with constant acceleration.

10–10 Using a base circle diameter of $4\frac{1}{2}$ in. and the displacement diagram of Problem 10–9, lay out the cam. The rotation is clockwise, and the follower is an in-line roller with $\frac{3}{8}$-in. diameter.

10–11 Construct the displacement diagram of a follower with simple harmonic motion. The follower rises 3 cm from 0 to 120°, dwells from 120 to 170°, falls 3 cm from 170 to 310°, and dwells from 310 to 360°. Knife edge 8$\frac{1}{2}$cm base circle.

10–12 Construct the displacement diagram of a follower having motion as follows: rise $1\frac{1}{2}$ in. from 0 to 150° with cycloidal motion, dwell from 150 to 185°, and fall $1\frac{1}{2}$ in. from 185 to 360° with cycloidal motion.

10–13 Construct the displacement diagram for the following motion: rise $1\frac{1}{2}$ in. from 0 to 170° with constant acceleration, dwell 170 to 200°, fall $1\frac{1}{2}$ in. from 200 to 360° with cycloidal motion. (NOTE: Use the right end of the diagram for cycloidal construction.) 3" base Knife edge

10–14 Plot the rise portion of a follower with simple harmonic motion and a 2-in. rise from 0 to 120°. Use six cam angle intervals, and measure the displacement at each interval. Using the mathematical method in Section 10–8, calculate the displacement for each interval and compare the results with the graphical results.

10–15 Do Problem 10–13 using cycloidal motion instead of simple harmonic motion and compare the results.

10–16 A cam design is needed in which the follower has zero acceleration during the middle portion of its travel. What type of motion should be used?

10–17 Using the displacement diagram of Problem 10–11, lay out a cam with a base circle diameter of 8.5 cm. The rotation is counter-clockwise, and the follower is an in-line flat-faced follower.

10–18 Use the displacement diagram of Problem 10–12 to lay out a cam with an offset roller follower. The follower is offset $\frac{3}{8}$ in. to the left, and the roller diameter is $\frac{1}{2}$ in. The base circle diameter is 4 in., and the cam rotates counterclockwise.

10–19 Lay out the same cam as in Problem 10–18, except change the base circle diameter to 2 in. Check the cam contour for maximum pressure angle and also undercutting. Do both these conditions appear acceptable?

10–20 Lay out a cam with oscillating roller follower, using the displacement diagram of Problem 10–12. The arm is to the right and has a length of $3\frac{1}{2}$ in. The base circle of the cam is 4 in., and it rotates counterclockwise.

CHAPTER 11
GEARING

11-1 INTRODUCTION

In this and following chapters we introduce a class of mechanisms completely different from any previously discussed, yet at the same time one of the most commonly used. *Gearing* is used in so many mechanisms that it is almost impossible to list them all. As noted in Chapter 1, the history of gearing is old, and some of Leonardo da Vinci's writings contained notes and sketches on gearing. Indeed, da Vinci developed the idea of a speed changer by substituting one gear for another.

We may define a gear as a toothed machine element that transmits motion to another element by engaging the teeth on one gear to those on a second gear. The study of gearing can be divided into three separate phases. One phase is the tooth profile, its development, and its action as the gear rotates. A second phase is the consideration of the geometric and kinematic equivalent of the gear, which allows us to calculate angular velocities and speed ratios. The third phase consists of the study of gear *trains*, where groups of gears are designed into a *train* to transmit power at certain speeds.

11-2 CLASSIFICATION OF GEARS

Somewhat like cams, there are different ways to classify gears. One method of classification is to consider the relationship of the shaft axes of the gears. Consider the shaft axes shown in Figure 11–1, which is the gear train and crankshaft of a Pratt and Whitney Twin Wasp aircraft radial engine. The engine is no longer manufactured, but the illustration presents a good view of the shafts on which the gears are mounted. An examination of these shafts shows parallel shaft axes for some gears, intersecting axes for other gears, and nonparallel and nonintersecting axes for other gears. This last

179

FIGURE 11–1. Gear train and crankshaft of Pratt and Whitney Twin *Wasp* engine (Courtesy United Technologies Corporation)

condition is referred to as *skewed*. This text classifies gearing by this means. We may then present the following summary of this method of classifying gears.

Gears may be classified by the relationship of the shafts on which they are mounted. The three shaft relationships are

1. **Parallel shafts.**
2. **Intersecting shafts.**
3. **Skewed shafts.**

The various gear *types* that we shall discuss fall into one or more of these classes. In most cases a specific gear type falls into only one class, but some gear types may be used with more than one class.

Gears may also be classified as *external* and *internal*. An *external* gear has teeth on the outside diameter; the *internal* gear has teeth on the inside diameter of the gear. Figure 11–2 illustrates the use of external and internal gearing. It is apparent in Figure

FIGURE 11–2. External and internal gears.

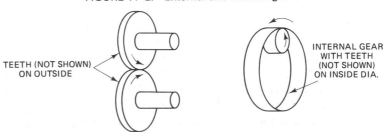

TEETH (NOT SHOWN) ON OUTSIDE

INTERNAL GEAR WITH TEETH (NOT SHOWN) ON INSIDE DIA.

(a) External gear   (b) Internal gear mated with external gear

11–2(b) that the large internal gear can only mesh with an external gear. Although external gearing is more common, internal gears are used in many applications. As with the shaft classification, some gear types may exist in both external and internal configurations.

11–3 GEAR TYPES

Gears are typed generally by considering their geometric form and the tooth configuration used. Both *spur gears* and *helical gears* take the form of a cylinder; the difference in tooth direction between the two constitutes the difference in type.

The *spur gear* in Figure 11–3 is cylindrical and has tooth elements running *parallel* to the gear axis. In Figure 11–4 the spur gear is called the *pinion*. *A pinion is the smaller of any two gears in mesh*. The pinion here is meshed with a *rack*, and the combination is called a *rack and pinion*. The rack may be thought of as another spur gear with an infinitely large radius.

The *helical gears* in Figure 11–5 are cylindrical in shape, but the teeth are not parallel to the axis of the gear. They are *skewed*

FIGURE 11–3 (left). Spur gear. (Courtesy Browning Manufacturing Division, Emerson Electric Co.)

FIGURE 11–5. Helical gears. (Courtesy Browning Manufacturing Division, Emerson Electric Co.)

FIGURE 11–4. The spur gear is the pinion of this rack and pinion combination. (Courtesy Browning Manufacturing Division, Emerson Electric Co.)

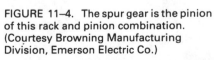

FIGURE 11–6. Bevel gears. (Courtesy Browning Manufacturing Division, Emerson Electric Co.)

FIGURE 11–7. Spiral bevel gears. (Courtesy Browning Manufacturing Division, Emerson Electric Co.)

with respect to the gear axis. Thus, we can say that a helical gear has the shape of a cylinder, and that its teeth are at some angle with respect to the gear axis.

The *bevel gears* in Figure 11–6 have the geometric form of cones. The teeth on the gears are straight and can be considered to be elements on a conical surface that intersect at the apex of the cone.

The *spiral bevel gears* in Figure 11–7 are shaped as cones, but the teeth are curved, making the gears different from the bevel gears of Figure 11–6.

FIGURE 11–8. Worm gear reducer. (Courtesy Winsmith Division of UMC Industries, Inc.)

FIGURE 11–9. Helical gears for parallel shaft mounting. (Courtesy Browning Manufacturing Division, Emerson Electric Co.)

182

A special type of gearing is the *worm and gear* combination of Figure 11–8. The *worm* most closely approximates the helix of a screw thread, and the tooth form of the gear is usually adapted to the worm.

The *types* of gears discussed here can be considered as primary types. Within a type there are usually one or more secondary types; these will be discussed later as we study each type more fully.

If we study Figures 11–3 through 11–8, we can visualize the relationship of the shafts on which the gears are mounted. When this is done, we can classify each type of gear by its shaft relationship, as described in Section 11–2. This is done in Table 11–1.

TABLE 11–1. Gear Type Classification by Shaft Relationship

Shaft Relationship	Gear Types Used
Parallel	Spur and helical
Intersecting	Bevel
Skew	Helical and worm

Note that helical gears are listed as being used for both parallel and skew shafts, although Figure 11–5 shows only the skew shaft condition. Two helical gears with parallel axes are shown in Figure 11–9. Explanations for the possible two-shaft conditions for helical gears are given in Chapter 13.

11–4 GEAR TOOTH ACTION

The profile of the gear tooth has to conform to definite criteria to provide smooth gear rotation at a constant angular velocity.

These criteria are contained in the fundamental law of gearing. It states the following:

The common normal at the point of contact between two teeth must always pass through a fixed point on the line connecting the centers of the two gears. This fixed point is called the *pitch point*.

The fundamental law of gearing can be explained by examining Figure 11–10. Here two teeth of a pair of gears are shown in contact. The common normal to the teeth at their contact passes through the *pitch point*, a point on the line connecting the centers of the two

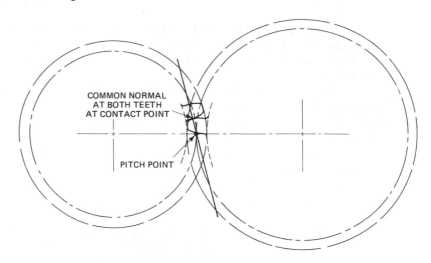

FIGURE 11–10. Common normal passing through pitch point.

FIGURE 11–11. Gear terms.

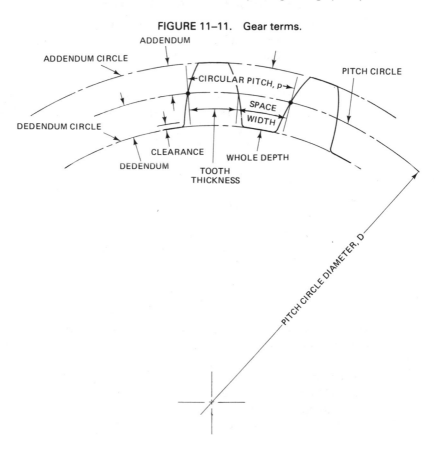

gears. Furthermore, the point of contact of these same two teeth *in any other position* will be on this same normal. Both the cycloid and involute curves give tooth profiles that satisfy the fundamental law of gearing.

Before we can proceed further with tooth profile development, it is necessary to define basic gear terms. Figure 11–11 shows a partial drawing of a spur gear. The gear teeth have the *involute* tooth form, which is used almost exclusively and is the only form that we shall consider in this text.

It will be noted in Figure 11–11 that most of the dimensions identified are measured from or along the pitch circle. The *pitch circle* is the theoretical circle upon which the kinematic analysis of gearing is based. Two mating spur gears are said to be kinematically equivalent to two cylinders rolling together without slip, as shown in Figure 11–12. The pitch circle represents the surface in contact with the mating gear, and the *pitch point* is located on the line of centers of the gears at the contact point of the two pitch circles.

Returning again to Figure 11–11, we can now define the other terms of the gear nomenclature as follows:

Know **Addendum.** The radial distance from the pitch circle to the outside diameter of the gear.

Addendum circle. The circle drawn through the outer end of the addendum. It corresponds to the circle determined by the outside diameter.

Know **Dedendum.** The radial distance from the pitch circle to the bottom of the tooth.

Dedendum circle. The circle drawn through the inner end of the dedendum.

Pitch circle diameter. The diameter of the pitch circle.

Know
Clearance

FIGURE 11–12. Two cylinders rolling without slip are the kinematic equivalent of two spur gears.

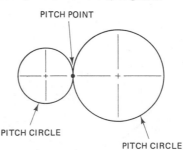

PITCH POINT

PITCH CIRCLE

PITCH CIRCLE

Circular pitch. **The distance measured along the pitch circle from a point on a tooth to the corresponding point on an adjacent tooth. Note that this distance is an** *arc.*

Tooth thickness. **The distance across the tooth measured along the pitch circle. It is an** *arc.*

Space width. **The distance measured along the pitch circle between two adjacent teeth. It is an** *arc.*

Know *Clearance.* **The distance between the bottom of one tooth and the top of the engaging tooth on the other gear. It is measured radially and is equal to the addendum subtracted from the dedendum.**

Whole depth. **The sum of the addendum and dedendum.**

The *involute* is the path traced by a point on a string as the string is unwound from a circle. It is shown in Figure 11–13. The *base circle* is the circle from which the string is unwound. The string forms a line tangent to the circle as it is unwound, and the location of its end is equal to the length of arc from its starting position. Thus, in Figure 11–13, the length of line *op'* is the arc length *oa*. The construction of the involute is made by plotting points on the tangents as shown.

We have said that teeth make contact along a line that is normal to the tooth profile at the contact point and passes through the pitch point. This common normal is called the *line of action* of the two gears. For two gears in contact, the *line of action* is also the line tangent to the base circles of the two gears, as in the con-

FIGURE 11–13. Involute construction.

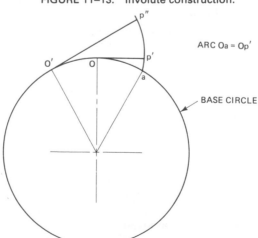

ARC Oa = Op'

BASE CIRCLE

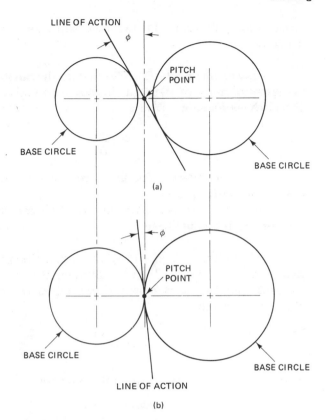

FIGURE 11–14. Line of action and base circle construction for two different angles ϕ.

struction of Figure 11–14, which shows the line of action for two pairs of gears having the same distance between centers and the same pitch diameters. The line of action makes an angle ϕ (phi), with a perpendicular to the line connecting the centers, when the perpendicular is constructed at the pitch point. The angle ϕ is called the *pressure angle*. Figure 11–14(a) and (b) have been drawn with different values of the pressure angle ϕ to show the change in the base circle diameters that results when ϕ is changed. In practice, the pressure angles of gears have been standardized at $14\frac{1}{2}$, 20, and 25°.

11–5 MATHEMATICAL RELATIONSHIPS

We can now use the information presented in the preceding sections to develop the mathematical relationships that are necessary in the selection and design of gears and gear trains. To do this, we use the terms shown in Figure 11–11 along with the additional terms defined here.

know **Diametral pitch, P.** **The ratio of the number of teeth to the pitch diameter of the gear. Letting n = number of teeth and D = pitch diameter in inches,**

$$P = \frac{n}{D}$$ (11–1)

Line of centers. **The line connecting the centers of rotation of two mating gears.**

Center distance. **The distance between the centers of rotation of two mating gears.**

Returning to Figure 11–11, the circumference of the pitch circle is equal to the pitch diameter, D, multiplied by π. Dividing the circumference by the number of teeth gives the circular pitch. Mathematically, this is

$$p = \frac{\pi D}{n}$$ (11–2)

where p = circular pitch in inches

n = number of teeth

This can be rearranged as follows:

$$p = \frac{\pi}{n/D}$$

But $n/D = P$, so that the equation becomes

$$p = \frac{\pi}{P}$$

or $$pP = \pi$$ (11–3)

Diametral pitch is one of the most important parameters of gear design and selection. Its importance can be summed up in the following rule.

For two gears to mesh, the diametral pitch of both gears must be the same.

The diametral pitches most commonly used include the following: 2, 3, 4, 6, 8, 10, 12, 16, 20, 24, and 48. This list is not inclusive, and other pitches are also used.

An equation for the *center distance* can be developed from Figure 11–15. For the pair of external gears shown in Figure 11–15(a), the equation for the center distance is

$$L_c = \frac{D_p}{2} + \frac{D_g}{2} \qquad (11\text{–}4)$$

where L_c = center distance in inches
          D_p = pitch diameter of the pinion (smaller gear) in inches
          D_g = pitch diameter of the gear in inches

For the internal gears shown in Figure 11–15(b), the equation is

$$L_c = \frac{D_g}{2} - \frac{D_P}{2} \qquad (11\text{–}5)$$

Equation 11–4 can be used to develop another useful form when the numbers of teeth of the two mating gears are known. This is done by a rearrangement of Equation 11–1 and substitution in Equation 11–4. Equation 11–1 is

$$P = \frac{n}{D}$$

Rearranging,

$$D = \frac{n}{P}$$

Substituting this in Equation 11–4, we have

$$L_c = \frac{n_p}{2P} + \frac{n_g}{2P} \qquad (11\text{–}6)$$

FIGURE 11–15. Center distance.

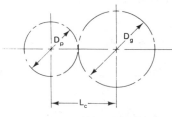

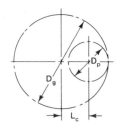

(a) External: $L_c = \dfrac{D_p}{2} + \dfrac{D_g}{2}$ (b) Internal: $L_c = \dfrac{D_g}{2} - \dfrac{D_p}{2}$

The subscripts p and g in the preceding equation indicate the pinion and the gear. The diametral pitch, P, is the same for both gears, since mating gears must have the same diametral pitch. A similar equation could be developed from Equation 11–5.

As pointed out before, spur gears and helical gears are kinematically equivalent to cylinders rolling in contact with each other, with the contact being on the pitch circle of each gear. This condition allows us to set equations for the gear angular velocities and for the *gear ratios*. Consider the two mating gears of Figure 11–16. The two cylinders roll without slip, and point A is the contact point. If there is no slip, then the linear velocity of A on the pinion and A on the gear must be the same. Using Equation 4–15 ($v = \omega r$), we can then relate the linear velocity to the gear angular velocities as follows. For the pinion,

$$v_A = \omega_p r_p$$

and

$$\omega_p = \frac{v_A}{r_p} \tag{11–7}$$

For the gear,

$$v_A = \omega_g r_g$$

and

$$\omega_g = \frac{v_A}{r_g} \tag{11–8}$$

If Equation 11–8 is divided by Equation 11–7, we have

$$\frac{\omega_g}{\omega_p} = \frac{v_A/r_g}{v_A/r_p}$$

Simplifying,

$$\frac{\omega_g}{\omega_p} = \frac{r_p}{r_g} \tag{11–9}$$

The radius, r, is equal to the pitch diameter, D, divided by 2. When this is substituted in the equation, it becomes

$$\frac{\omega_g}{\omega_p} = \frac{D_p}{D_g} \tag{11–10}$$

Equation 11–10 is a very important equation. It can be stated as follows:

The angular velocity ratio of any two mating gears is equal to the *inverse* ratio of the pitch diameters of the gears.

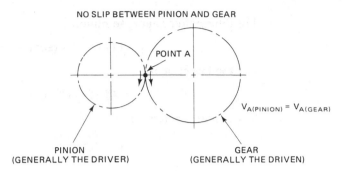

NO SLIP BETWEEN PINION AND GEAR

POINT A

$V_{A(PINION)} = V_{A(GEAR)}$

PINION
(GENERALLY THE DRIVER)

GEAR
(GENERALLY THE DRIVEN)

FIGURE 11–16.   Gear velocity.

If the term n/P (from Equation 11–1) is substituted for D in Equation 11–10, we have

$$\frac{\omega_g}{\omega_p} = \frac{n_p}{n_g} \qquad \text{(11–11)}$$

The angular velocity ratio of any two mating gears is equal to the *inverse* ratio of the number of teeth of the gears.

As noted in Figure 11–6, the pinion is generally the *driver* and the gear is the *driven* member of the pair. With this relationship and with the equations developed here, the gears in Figure 11–16 function as a *gear reducer*. The velocity ratio, ω_g/ω_p, for this condition is always less than 1. The pinion is not the driver in all cases, however, and the reader should be careful to note any exceptions.

The equations developed here are important general equations applicable to all gears, which are kinematically equivalent to cylinders. Their use is shown in the following examples.

EXAMPLE
A pair of mating gears has a diametral pitch of 4 and pitch diameters of 6 and 8 in. Determine the number of teeth on each gear, the center distance, and the velocity ratio.

Solution
The number of teeth is determined using Equation 11–1. Rearranging the equation, we have

$$n = PD$$

For the smaller gear (the pinion),

$$n_p = (4)(6) = 24 \text{ teeth}$$

For the larger gear,

$$n_g = (4)(8) = 32 \text{ teeth}$$

The center distance is

$$L = \frac{D_p}{2} + \frac{D_g}{2}$$

$$= \frac{6}{2} + \frac{8}{2}$$

$$= 7 \text{ in.}$$

The velocity ratio is found from Equation 11–10.

$$\frac{\omega_g}{\omega_p} = \frac{D_p}{D_g}$$

$$= \frac{6}{8} = 0.75$$

EXAMPLE

A gear with 36 teeth and a diametral pitch of 3 is to be used as the pinion of a pair of gears with a desired velocity ratio of 0.71. Determine the number of teeth of the mating gear, the center distance, and the pitch diameter of both gears.

Solution

Using Equation 11–1, we can find the pitch diameter of the first gear.

$$D = \frac{n}{P} = \frac{36}{3}$$

$$= 12 \text{ in.}$$

The desired velocity ratio, ω_g/ω_p, is 0.71. From Equation 11–11, this also is the ratio of the number of teeth, n_p/n_g. Thus,

$$\frac{\omega_g}{\omega_p} = 0.71 = \frac{n_p}{n_g}$$

Then

$$0.71 = \frac{36}{n_g}$$

Solving for n_g,

$$n_g = \frac{36}{0.71} = 50.7 \text{ teeth}$$

A gear with 50.7 teeth is obviously impossible. It is therefore necessary to select the nearest whole number for the number of teeth. In this case it is 51 teeth. The *actual* velocity ratio with 51 teeth is 36/51, or 0.706.

The pitch diameter of the gear is

$$D_g = \frac{n_g}{P} = \frac{51}{3}$$

$$= 17 \text{ in.}$$

The center distance is

$$L_c = \frac{12}{2} + \frac{17}{2}$$

$$= 14\tfrac{1}{2} \text{ in.}$$

EXAMPLE

Two mating spur gears have 26 teeth and 74 teeth. The pinion rotates at 1250 rpm. What is the rotational speed of the gear?

Solution

From Equation 11–11, the angular velocity ratio is

$$\frac{\omega_g}{\omega_p} = \frac{n_p}{n_g} = \frac{26}{74}$$

The angular velocity, ω, is in *radians* per unit time. However, since we are dealing with a ratio, the conversion using 2π cancels out. We can then use revolutions per minute directly and solve for ω_g, as follows.

$$\omega_g = \frac{26}{74}(\omega_p) = \frac{26}{74}(1250)$$

$$= 439 \text{ rpm}$$

11–6   OTHER GEAR TERMS

Figure 11–17 illustrates some of the gear surfaces and gives the names of these surfaces. The *flank* of the gear tooth is important in wear considerations, since sliding action caused by the teeth of the mating gear occurs on the flank.

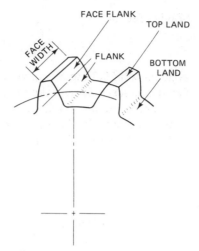

FIGURE 11–17. Gear parts.

FIGURE 11–18. Gear tooth interference.   FIGURE 11–19. Gear backlash.

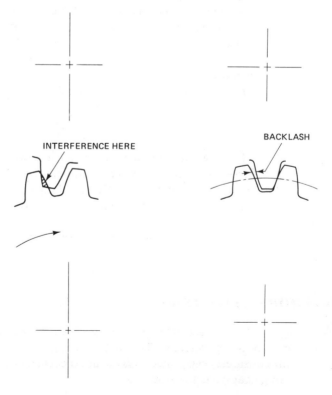

Interference of gear teeth can occur under certain conditions. An example of interference is shown in Figure 11–18. Interference will be taken up in detail in later chapters.

Gear backlash is the difference between the tooth thickness and the space width, as shown in Figure 11–19. Zero backlash would exist theoretically if the tooth thickness and the space width were equal. In practice, the space width is made slightly larger, since manufacturing inaccuracies prevent achieving zero backlash.

► **PROBLEMS**

11–1 Name two types of gears that are used for parallel shafts.

11–2 What gear type is used for intersecting shafts?

11–3 What two gear types are used for skew shafts?

11–4 Define the involute.

11–5 A spur gear has a diametral pitch of 4. What is the circular pitch?

11–6 For two gears to mesh, the _____ of both gears has to be the same. (Fill in the correct answer.)

11–7 An internal gear meshes with a spur gear. The pitch diameter is 12 in. for the internal gear and 6 in. for the spur gear. Find the center distance.

11–8 A spur gear has a diametral pitch of 4 and 32 teeth. Find the pitch diameter and the circular pitch.

11–9 A pinion has a pitch diameter of 3.5 in. and a diametral pitch of 8. It is to mate with a gear so that the velocity ratio is $\frac{1}{2}$. Find (a) the number of teeth on the gear, (b) the diametral pitch of the gear, (c) the pitch diameter of the gear, and (d) the center distance.

11–10 A gear mates with a pinion. The gear has 105 teeth and the pinion 35 teeth. If the gear rotates at 265 rpm, what is the speed of the pinion?

11–11 Find the velocity ratio for a pinion and gear having a 10-in. center distance. The diametral pitch is 6 and the pitch diameter of the pinion is 3 in.

11–12 The circular pitch of a gear is 0.5236 in. and its pitch diameter is 6 in. Find the diametral pitch and the number of teeth.

11–13 Name two of the three standard pressure angles.

11–14 The pinion of two mating gears rotates at 1200 rpm and the output gear rotates at 350 rpm. The diametral pitch is 4 and the pinion has 28 teeth. Find (a) the pitch diameters of both gears, (b) the number of teeth on the larger gear, and (c) the center distance.

11–15   Two gears have 26 and 72 teeth, respectively. The diametral pitch is 2. What is the center distance and the velocity ratio?

11–16   Two gears have speeds in the ratio of 1 to 1.5. Using a diametral pitch of 4, find the number of teeth on each gear if the center distance is 10 in.

CHAPTER 12
SPUR GEARS

12-1 INTRODUCTION

The material given in Chapter 11 on gears is broadly applicable to all types of gears. However, as we take up the other types of gears, in some cases the definitions and mathematical models have to be modified to fit the gear type in question. No modification is necessary for the spur gear, which we consider in this chapter.

12-2 STANDARD SPUR GEARS

The American National Standards Institute (ANSI) has standardized certain forms of the involute tooth form for spur gears. These standards are available in publications, and the standard ANSI B6.1-1968 gives data on involute tooth forms with 20 and 25° pressure angles. The older $14\frac{1}{2}$° full depth involute tooth system, although still available, is now considered obsolescent and is not an ANSI standard. The ANSI standard for the 20 and 25° tooth forms includes a *coarse pitch* (*pitch* here refers to *diametral pitch*) for pitches less than 20 and a *fine pitch* for pitches of 20 and greater. A summary of the 20 and 25° tooth form specifications, along with the older $14\frac{1}{2}$°, is presented in Table 12–1. The minimum allowable number of teeth is that which has been found to give satisfactory tooth action without interference. Although based on the mating of a spur and a rack, it also applies when two spur gears are mated.

12-3 INTERFERENCE

Interference of two teeth was shown in Figure 11–18. Interference occurs when two gear teeth contact outside the base circle, as shown in Figure 12–1. The initial contact point A, called the *interference*

TABLE 12–1.   Specifications for $14\frac{1}{2}$, 20, and 25° Tooth Forms

	$14\frac{1}{2}°$ Full Depth Involute (Obsolescent)	20 and 25° Involute Coarse (Less than 20 P)	20 and 25° Involute Fine (20 P and Greater)
Addendum	$\dfrac{1.000}{P}$	$\dfrac{1.000}{P}$	$\dfrac{1.000}{P}$
Dedendum	$\dfrac{1.157}{P}$	$\dfrac{1.250}{P}$	$\dfrac{1.200}{P}+0.002$
Clearance	$\dfrac{0.157}{P}$	$\dfrac{0.250}{P}$	$\dfrac{0.200}{P}+0.002$
Minimum allowable no. of teeth	32	20°: 18 25°: 12	Not applicable

point, is outside the base circle of the pinion centered at o_1. The interfering tooth causes the *undercut* shown in Figure 12–1(b). In fact, gear teeth are sometimes machined to this undercut form to prevent interference. It is obvious that this method weakens the gear tooth, however. Reducing the addendum by the amount of interference also would prevent interference. In Figure 12–1(a), the shaded tooth area would be removed to accomplish this. Using the minimum number of teeth as shown in Table 12–1 provides assurance that there will be no interference.

A check for the presence of interference can be done graphically as long as the pitch diameters, pressure angle, and addendums of the two gears are known. An example follows.

EXAMPLE
A $3\frac{1}{2}$-in.-pitch-diameter gear is to be used with a 4-in.-diameter gear. The diametral pitch is 4 and the pressure angle is 20°. Check to see if interference occurs.

Solution
Referring to Table 12–1, the addendum for a 20° pressure angle gear is $1.000/P$, or $\frac{1}{4}$, for the gears in question.

The line of centers, with the gear centers and pitch point, is constructed as shown in Figure 12–2. The pressure angle is then constructed. Points A and B are then located from perpendiculars drawn to the gear centers. The addendum of $\frac{1}{4}$ in. is then added to each pitch diameter, and

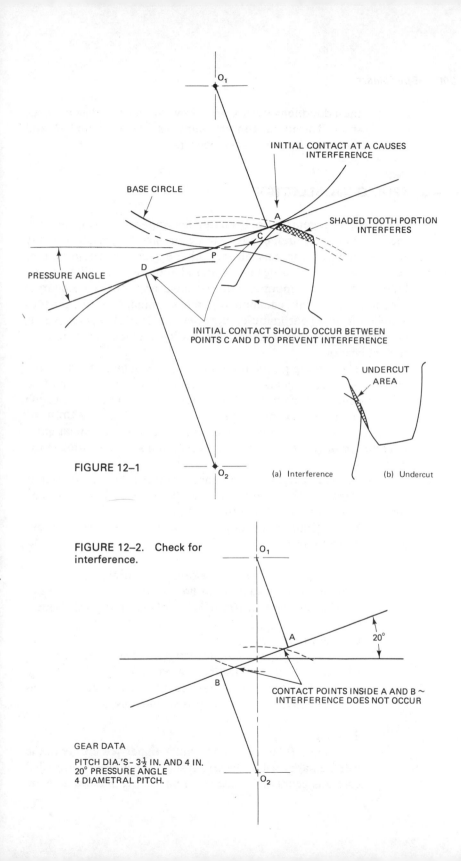

INITIAL CONTACT AT A CAUSES
INTERFERENCE

BASE CIRCLE

A

SHADED TOOTH PORTION
INTERFERES

C

P

D

PRESSURE ANGLE

INITIAL CONTACT SHOULD OCCUR BETWEEN
POINTS C AND D TO PREVENT INTERFERENCE

UNDERCUT
AREA

FIGURE 12–1

O_2

(a) Interference

(b) Undercut

FIGURE 12–2. Check for
interference.

O_1

A

20°

B

CONTACT POINTS INSIDE A AND B ~
INTERFERENCE DOES NOT OCCUR

GEAR DATA

PITCH DIA.'S – $3\frac{1}{2}$ IN. AND 4 IN.
20° PRESSURE ANGLE
4 DIAMETRAL PITCH.

O_2

the addendum circles are drawn in, as the dashed lines show. The initial contact points are inside A and B, and therefore no interference occurs.

12–4 SPUR GEAR SELECTION

The selection of gears for a specific engineering application includes the kinematic requirement that the gears have a certain velocity ratio. In addition, there usually are other engineering requirements relating to size of the gears, material requirements dictated by loading, method of mounting on the shaft, and other gear characteristics. It is usually cheaper to purchase standard stock gears as supplied from a gear manufacturer than to fabricate special gears, and in most cases these standard gears will meet the requirements of the application.

The catalog pages in Figures 12–3 through 12–7 show the availability of spur gears of diametral pitches 4, 6, 7, 12, and 16 from one gear manufacturer, the Browning Manufacturing Division of the Emerson Electric Co. Gear selection usually is done by referring to a catalog such as this. The examples of typical gear-selection problems that follow use these catalog pages to select the gears.

For the gear selection made here, only the kinematic requirements of the application are used. Other engineering requirements relating to materials and the like, are not considered.

Before using the catalog, we review the requirements for gears that are to mesh:

1. **The gears must have the same diametral pitch.**
2. **The pressure angle must be the same.**
3. **The gear system (from Table 12–1) must be the same.**

EXAMPLE
Select two gears of 16 diametral pitch that when mounted, will have a center distance of exactly 8 in. The pinion rotates at 1200 rpm and the output speed is to be as close as possible to 850 rpm. The pressure angle is 20°.

Solution
The diametral pitch is shown in the upper-left corner of the catalog pages and is described simply as pitch. Figure 12–3 contains gears of 16 diametral pitch, with pitch diameters

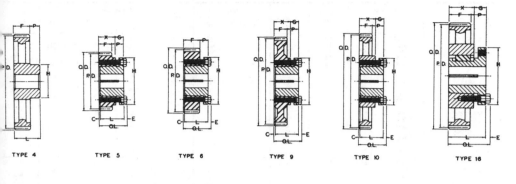

TYPE 4   TYPE 5   TYPE 6   TYPE 9   TYPE 10   TYPE 16

Table No. 1 — Stock Steel and Cast Iron Minimum Bore Spur Gears

Part No.	Diameter Pitch	Diameter Nominal O.D.	No. Teeth	Type	Bore Stock	Bore Max.	F	L	P	H	Wt. Lbs.
YSS1612	.750"	.87"	12	2	3/8"	3/8"	3/4"	1 1/4"	1/2"	9/16"	.1
YSS1614	.875	1.00	14	2	3/8	7/16	3/4	1 1/4	1/2	11/16	.1
YSS1615	.938	1.06	15	2	3/8	1/2	3/4	1 1/4	1/2	3/4	.1
YSS1616	1.000	1.12	16	2	1/2	1/2	3/4	1 1/4	1/2	13/16	.1
YSS1618	1.125	1.25	18	2	1/2	9/16	3/4	1 1/4	1/2	15/16	.2
YSS1620	1.250	1.37	20	2	5/8	5/8	3/4	1 1/4	1/2	1 1/16	.3
YSS1624	1.500	1.62	24	2	5/8	3/4	3/4	1 1/4	1/2	1 5/16	.4
YSS1628	1.750	1.87	28	2	5/8	7/8	3/4	1 1/4	1/2	1 1/2	.6
YSS1630	1.875	2.00	30	2	5/8	1	3/4	1 1/4	1/2	1 5/8	.7
YSS1632	2.000	2.12	32	2	5/8	1	3/4	1 1/4	1/2	1 3/4	.8
YSS1636	2.250	2.37	36	2	5/8	1 1/4	3/4	1 1/4	1/2	2	1.1
YSS1640	2.500	2.62	40	2	5/8	1 3/8	3/4	1 3/8	5/8	2 1/4	1.5
YSS1648	3.000	3.12	48	2	5/8	1 3/4	3/4	1 3/8	5/8	2 1/4	2.2
YSS1656	3.500	3.62	56	2	5/8	1 3/8	3/4	1 3/8	5/8	2 1/4	2.6
YSS1660	3.750	3.87	60	2	3/4	1 3/8	3/4	1 3/8	5/8	2 1/4	2.9
YSS1664	4.000	4.12	64	2	3/4	1 3/8	3/4	1 1/2	3/4	2 1/4	3.3
YSS1672	4.500	4.62	72	3	3/4	1 5/8	3/4	1 1/2	3/4	2 1/2	3.7
YSS1680	5.000	5.12	80	3	3/4	1 7/8	3/4	1 1/2	3/4	2 1/2	4.2
YCS1696	6.000	6.12	96	4	3/4	1 1/2	3/4	1 1/2	3/4	2 1/2	3.7
YCS16112	7.000	7.12	112	4	3/4	1 1/2	3/4	1 1/2	3/4	2 1/2	4.3
YCS16128	8.000	8.12	128	4	3/4	1 1/2	3/4	1 1/2	3/4	2 1/2	5.0
YCS16144	9.000	9.12	144	4	3/4	1 1/2	3/4	1 1/2	3/4	2 1/2	5.4
YCS16160	10.000	10.12	160	4	7/8	1 1/2	3/4	1 1/2	3/4	2 1/2	6.0
YCS16192	12.000	12.12	192	4	7/8	1 1/2	3/4	1 3/4	1	2 1/2	7.9

All 80 Tooth Gears and smaller are Steel. All larger sizes, Cast Iron.

Table No. 2 — Stock Steel and Cast Iron Spur Gears with Split Taper Bushings

Part No. Gear	Bush.	Diameter Pitch	Diameter Nominal O.D.	No. Teeth	Type	F	O.L.	L	P	C	H	G	X	E	Wt. Lbs. Less Bush.
'SS16G36	G	2.250"	2.37"	36	6	3/4"	1 3/8"	1"	7/16"	3/16"	2"	—	—	3/16"	.6
'SS16G40	G	2.500	2.62	40	6	3/4	1 3/8	1	7/16	3/16	2	—	—	3/16	.8
'SS16H48	H	3.000	3.12	48	5	3/4	1 1/2	1 1/4	9/16	1/16	2 1/2	7/16	7/8"	3/16	1.0
'SS16H56	H	3.500	3.62	56	5	3/4	1 1/2	1 1/4	9/16	1/16	2 1/2	7/16	7/8	3/16	1.7
'SS16H60 ■	H	3.750	3.87	60	5	3/4	1 1/2	1 1/4	9/16	1/16	2 1/2	7/16	7/8	3/16	2.0
'SS16H64	H	4.000	4.12	64	5	3/4	1 1/2	1 1/4	9/16	1/16	2 1/2	7/16	7/8	3/16	2.3
'SS16H72 ■	H	4.500	4.62	72	5	3/4	1 1/2	1 1/4	9/16	1/16	2 1/2	7/16	7/8	3/16	3.0
'SS16H80 ■	H	5.000	5.12	80	9	3/4	1 1/2	1 1/4	9/16	1/16	2 1/2	7/16	7/8	3/16	3.0
CS16H96	H	6.000	6.12	96	10	3/4	1 1/2	1 1/4	9/16	1/16	2 1/2	7/16	7/8	3/16	3.0
CS16H112 ■	H	7.000	7.12	112	10	3/4	1 1/2	1 1/4	9/16	1/16	2 1/2	7/16	7/8	3/16	3.7
CS16H128 ■	H	8.000	8.12	128	10	3/4	1 1/2	1 1/4	9/16	1/16	2 1/2	7/16	7/8	3/16	4.4
CS16H144 ■	H	9.000	9.12	144	10	3/4	1 1/2	1 1/4	9/16	1/16	2 1/2	7/16	7/8	3/16	4.8
CS16H160 ■	H	10.000	10.12	160	10	3/4	1 1/2	1 1/4	9/16	1/16	2 1/2	7/16	7/8	3/16	5.4
CS16P192 ■	P1	12.000	12.12	192	16	3/4	2 3/16	1 15/16	1 3/16	—	3	7/8	1 5/16	1/4	8.0

All 80 Tooth Gears and smaller are Steel. All larger sizes, Cast Iron.
■ Will not be restocked; available in production quantities only when present supply is exhausted.

FIGURE 12–3.   Spur gears: 16 pitch, 3/4-in. face, 20° pressure angle. (Courtesy Browning Manufacturing Division, Emerson Electric Co.)

ranging from 0.750 to 12.000 in. The problem then is to find a combination of gears from this page that most nearly meets the requirements.

Stock Steel and Cast Iron Minimum Bore Spur Gears

Part No.	Diameter		No. Teeth	Type	Bore		Dimensions				Wt. Lbs.
	Pitch	Nominal O.D.			Stock	Max.	F	L	P	H	
YSS1212	1.000"	1.16"	12	2	1/2"	1/2"	1"	1 5/8"	5/8"	3/4"	.2
YSS1213	1.083	1.25	13	2	5/8	5/8	1	1 5/8	5/8	13/16	.2
YSS1214	1.167	1.33	14	2	5/8	5/8	1	1 5/8	5/8	29/32	.2
YSS1215	1.250	1.41	15	2	5/8	5/8	1	1 5/8	5/8	1	.3
YSS1216	1.333	1.50	16	2	5/8	5/8	1	1 5/8	5/8	1 1/16	.4
YSS1218	1.500	1.66	18	2	3/4	3/4	1	1 5/8	5/8	1 1/4	.4
YSS1220	1.667	1.83	20	2	3/4	3/4	1	1 5/8	5/8	1 13/32	.6
YSS1221	1.750	1.91	21	2	3/4	3/4	1	1 5/8	5/8	1 1/2	.7
YSS1224	2.000	2.16	24	2	3/4	1	1	1 5/8	5/8	1 3/4	1.0
YSS1228	2.333	2.50	28	2	3/4	1 1/4	1	1 7/8	5/8	2 1/16	1.4
YSS1230	2.500	2.66	30	2	3/4	1 3/8	1	1 7/8	5/8	2 1/4	1.8
YSS1236	3.000	3.16	36	2	3/4	1 5/8	1	1 7/8	7/8	2 1/2	3.0
YSS1242	3.500	3.66	42	2	3/4	1 5/8	1	1 7/8	7/8	2 1/2	3.6
YSS1248	4.000	4.16	48	2	7/8	1 5/8	1	1 7/8	7/8	2 1/2	4.5
YSS1254	4.500	4.66	54	2	7/8	1 5/8	1	1 7/8	7/8	2 1/2	5.4
YSS1260	5.000	5.16	60	3	7/8	1 5/8	1	1 7/8	7/8	2 1/2	5.2
YCS1266	5.500	5.66	66	4	7/8	1 9/16	1	1 7/8	7/8	2 5/8	4.6
YCS1272	6.000	6.16	72	4	7/8	1 9/16	1	1 7/8	7/8	2 5/8	5.0
YCS1284	7.000	7.16	84	4	7/8	1 9/16	1	1 7/8	7/8	2 5/8	5.8
YCS1296	8.000	8.16	96	4	7/8	1 9/16	1	1 7/8	7/8	2 5/8	6.6
YCS12108	9.000	9.16	108	4	7/8	1 9/16	1	1 7/8	7/8	2 5/8	7.1
YCS12120	10.000	10.16	120	4	1	1 9/16	1	1 7/8	7/8	2 5/8	7.8
YCS12132	11.000	11.16	132	4	1	1 9/16	1	2	1	2 5/8	10.9
YCS12144	12.000	12.16	144	4	1	1 9/16	1	2	1	2 5/8	11.8
YCS12168	14.000	14.16	168	4	1	1 9/16	1	2	1	2 5/8	13.7
YCS12192 ■	16.000	16.16	192	4	1	1 9/16	1	2	1	2 5/8	17.3
YCS12216 ■	18.000	18.16	216	4	1	1 5/8	1	2	1	2 3/4	21.0
YCS12240 ■	20.000	20.16	240	4	1	1 5/8	1	2	1	2 3/4	22.9

All 60 Tooth Gears and smaller are Steel. All larger sizes, Cast Iron.
■Will not be restocked; available in production quantities only when present supply is exhausted.

Stock Steel and Cast Iron Spur Gears with Split Taper Bushings

Part No.		Diameter		No. Teeth	Type	Dimensions									Wt. Lbs. Less Bush.
Gear	Bush.	Pitch	Nominal O.D.			F	O.L.	L	P	C	H	G	X	E	
YSS12G30 ■	G	2.500"	2.66"	30	6	1"	1 5/8"	1"	7/16"	7/16"	2"	—	—	3/16"	1.1
YSS12H36	H	3.000	3.16	36	6	1	1 5/8	1 1/4	7/16	3/16	2 1/2	—	—	3/16	1.4
YSS12H42	H	3.500	3.66	42	6	1	1 5/8	1 1/4	7/16	3/16	2 1/2	—	—	3/16	2.0
YSS12H48	H	4.000	4.16	48	6	1	1 5/8	1 1/4	7/16	3/16	2 1/2	—	—	3/16	3.0
YSS12H54	H	4.500	4.66	54	6	1	1 5/8	1 1/4	7/16	3/16	2 1/2	—	—	3/16	4.0
YSS12H60	H	5.000	5.16	60	7	1	1 5/8	1 1/4	7/16	3/16	2 1/2	—	—	3/16	3.5
YCS12H66 ■	H	5.500	5.66	66	8	1	1 1/2	1 1/4	5/16	1/16	2 1/2	7/16"	7/8"	3/16	3.3
YCS12H72	H	6.000	6.16	72	8	1	1 1/2	1 1/4	5/16	1/16	2 1/2	7/16	7/8	3/16	3.4
YCS12H84	H	7.000	7.16	84	8	1	1 1/2	1 1/4	5/16	1/16	2 1/2	7/16	7/8	3/16	4.5
YCS12H96	H	8.000	8.16	96	8	1	1 1/2	1 1/4	5/16	1/16	2 1/2	7/16	7/8	3/16	5.2
YCS12H108	H	9.000	9.16	108	8	1	1 1/2	1 1/4	5/16	1/16	2 1/2	7/16	7/8	3/16	5.8
YCS12H120 ■	H	10.000	10.16	120	8	1	1 1/2	1 1/4	5/16	1/16	2 1/2	7/16	7/8	3/16	6.6
YCS12P132 ■	P1	11.000	11.16	132	16	1	2 3/16	1 15/16	15/16	—	3	7/8	1 5/16	1/4	9.8
YCS12P144	P1	12.000	12.16	144	16	1	2 3/16	1 15/16	15/16	—	3	7/8	1 5/16	1/4	10.5
YCS12P168 ■	P1	14.000	14.16	168	16	1	2 3/16	1 15/16	15/16	—	3	5/8	1 5/16	1/4	13.0
YCS12P192 ■	P1	16.000	16.16	192	16	1	2 3/16	1 15/16	15/16	—	3	5/8	1 5/16	1/4	17.3
YCS12P216 ■	P1	18.000	18.16	216	16	1	2 3/16	1 15/16	15/16	—	3	5/8	1 5/16	1/4	20.3
YCS12P240 ■	P1	20.000	20.16	240	16	1	2 3/16	1 15/16	15/16	—	3	5/8	1 5/16	1/4	21.9

All 60 Tooth Gears and smaller are Steel. All larger sizes, Cast Iron.
■Will not be restocked; available in production quantities only when present supply is exhausted.

FIGURE 12–4. Spur gears: 12 pitch, 1-in. face, 20° pressure angle. (Courtesy Browning Manufacturing Division, Emerson Electric Co.

The equations involved are

$$L_c = \frac{D_p}{2} + \frac{D_g}{2} \qquad (11-4)$$

and

$$\frac{\omega_g}{\omega_p} = \frac{D_p}{D_g} \qquad (11-10)$$

Using Equation 11–4, the center distance is 8 in. Substituting this into the equation gives

Table No. 1 **Stock Steel and Cast Iron Minimum Bore Spur Gears**

Part No.	Diameter		No. Teeth	Type	Bore		Dimensions				Wt. Lbs.
	Pitch	Nominal O.D.			Stock	Max.	F	L	P	H	
YSS812	1.500"	1.75"	12	2	¾"	¾"	1½"	2¼"	¾"	1⅛"	.6
YSS814	1.750	2.00	14	2	¾	¾	1½	2¼	¾	1⅜	1.0
YSS815	1.875	2.12	15	2	¾	⅞	1½	2¼	¾	1½	1.2
YSS816	2.000	2.25	16	2	⅞	1	1½	2⅜	⅞	1⅝	1.4
YSS818	2.250	2.50	18	2	⅞	1⅛	1½	2⅜	⅞	1⅞	1.9
YSS820	2.500	2.75	20	2	⅞	1¼	1½	2⅜	⅞	2⅛	2.5
YSS822	2.750	3.00	22	2	⅞	1½	1½	2⅜	⅞	2⅜	3.0
YSS824	3.000	3.25	24	2	⅞	1¾	1½	2⅜	⅞	2⅝	3.9
YSS828	3.500	3.75	28	2	⅞	2⅛	1½	2⅜	⅞	3⅛	5.4
YSS832	4.000	4.25	32	2	1	2⅛	1½	2½	1	3⅛	6.9
YSS836	4.500	4.75	36	2	1	2⅛	1½	2½	1	3⅛	8.3
YSS840	5.000	5.25	40	2	1	2⅛	1½	2½	1	3⅛	9.9
YSS844	5.500	5.75	44	2	1	2¼	1½	2½	1	3¼	12.5
YSS848	6.000	6.25	48	2	1	2¼	1½	2½	1	3¼	14.5
YSS856	7.000	7.25	56	2	1	2¼	1½	2½	1	3¼	19.0
YSS860	7.500	7.75	60	2	1	2¼	1½	2½	1	3¼	21.1
YCS864	8.000	8.25	64	4	1	2	1½	2½	1	3¼	12.9
YCS872	9.000	9.25	72	4	1	2	1½	2½	1	3¼	14.3
YCS880	10.000	10.25	80	4	1⅛	2	1½	2¾	1¼	3¼	16.6
YCS888	11.000	11.25	88	4	1⅛	2	1½	2¾	1¼	3¼	17.8
YCS896	12.000	12.25	96	4	1⅛	2	1½	2¾	1¼	3¼	20.1
YCS8112	14.000	14.25	112	4	1⅛	2	1½	2¾	1¼	3¼	25.1
YCS8120 ■	15.000	15.25	120	4	1⅛	2	1½	2¾	1¼	3¼	26.7
YCS8128	16.000	16.25	128	4	1⅛	2	1½	2¾	1¼	3¼	28.6
YCS8144	18.000	18.25	144	4	1⅛	2	1½	2¾	1¼	3¼	34.9
YCS8160	20.000	20.25	160	4	1¼	2¼	1½	3	1½	3½	42.1
YCS8176 ■	22.000	22.25	176	4	1¼	2¼	1½	3	1½	3½	45.9
YCS8192 ■	24.000	24.25	192	4	1¼	2¼	1½	3	1½	3¾	52.0

All 60 Tooth Gears and Smaller are Steel. All Larger Sizes, Cast Iron.
■Will not be restocked; available in production quantities only when present supply is exhausted.

Table No. 2 **Stock Steel and Cast Iron Spur Gears with Split Taper Bushings**

Part No.		Diameter		No. Teeth	Type	Dimensions									Wt. Lbs. Less Bush.
Gear	Bush.	Pitch	Nominal O.D.			F	O. L.	L	P	C	H	G	X	E	
YSS8P28	P1	3.500"	3.75"	28	11	1½"	2⅜"	1 15/16"	⅝"	3/16"	3"	—	—	¼"	2.8
YSS8P32	P1	4.000	4.25	32	11	1½	2⅜	1 15/16	⅝	3/16	3	—	—	¼	3.9
YSS8P36	P1	4.500	4.75	36	12	1½	2 3/16	1 15/16	7/16	—	3	⅝"	1 5/16"	¼	5.1
YSS8P40	P1	5.000	5.25	40	12	1½	2 3/16	1 15/16	7/16	—	3	⅝	1 5/16	¼	6.4
YSS8P44	P1	5.500	5.75	44	12	1½	2 3/16	1 15/16	7/16	—	3	⅝	1 5/16	¼	8.0
YSS8P48	P1	6.000	6.25	48	12	1½	2 3/16	1 15/16	7/16	—	3	⅝	1 5/16	¼	10.1
YSS8P56	P1	7.000	7.25	56	12	1½	2 3/16	1 15/16	7/16	—	3	⅝	1 5/16	¼	14.0
YSS8P60	P1	7.500	7.75	60	12	1½	2 3/16	1 15/16	7/16	—	3	⅝	1 5/16	¼	15.8
YCS8P64	P1	8.000	8.25	64	17	1½	2 5/16	1 15/16	7/16	—	3	⅝	1 5/16	¼	9.8
YCS8P72	P1	9.000	9.25	72	17	1½	2 3/16	1 15/16	7/16	—	3	⅝	1 5/16	¼	10.9
YCS8P80	P1	10.000	10.25	80	17	1½	2 3/16	1 15/16	7/16	—	3	⅝	1 5/16	¼	12.3
YCS8P88■	P1	11.000	11.25	88	17	1½	2 3/16	1 15/16	7/16	—	3	⅝	1 5/16	¼	14.7
YCS8P96■	P1	12.000	12.25	96	17	1½	2 3/16	1 15/16	7/16	—	3	⅝	1 5/16	¼	16.9
YCS8Q112	Q1	14.000	14.25	112	16	1½	2 25/32	2½	1	—	4⅛	¾	1¾	9/32	21.9
YCS8Q120	Q1	15.000	15.25	120	16	1½	2 25/32	2½	1	—	4⅛	¾	1¾	9/32	24.8
YCS8Q128■	Q1	16.000	16.25	128	16	1½	2 25/32	2½	1	—	4⅛	¾	1¾	9/32	28.3
YCS8Q144■	Q1	18.000	18.25	144	16	1½	2 25/32	2½	1	—	4⅛	¾	1¾	9/32	31.4
YCS8Q160■	Q1	20.000	20.25	160	16	1½	2 25/32	2½	1	—	4⅛	¾	1¾	9/32	44.9
YCS8Q176■	Q1	22.000	22.25	176	16	1½	2 25/32	2½	1	—	4⅛	¾	1¾	9/32	39.4
YCS8Q192■	Q1	24.000	24.25	192	16	1½	2 25/32	2½	1	—	4⅛	¾	1¾	9/32	44.8

All 60 Tooth Gears and Smaller are Steel. All Larger Sizes, Cast Iron.
■Will not be restocked; available in production quantities only when present supply is exhausted.

FIGURE 12–5. Spur gears: 8 pitch, $1\frac{1}{2}$-in. face, 20° pressure angle. *Note*: 20° pressure angle gears will not mesh with $14\frac{1}{2}°$ pressure angle gears. (Courtesy Browning Manufacturing Division, Emerson Electric Co.)

$$8 = \frac{D_p}{2} + \frac{D_g}{2}$$

Rearranging,

$$D_p + D_g = (2)(8) = 16$$

The *desired* value of the velocity ratio is

$$\frac{\omega_g}{\omega_p} = \frac{850}{1200} = 0.708$$

Table No. 1 **Stock Steel and Cast Iron Minimum Bore Spur Gears**

Part No.	Diameter		No. Teeth	Type	Bore		Dimensions				Wt. Lbs.
	Pitch	Nominal O.D.			Stock	Max.	F	L	P	H	
YSS612	2.000"	2.33"	12	2	1"	1"	2"	2 7/8"	7/8"	1 1/2"	1.5
YSS614	2.333	2.66	14	2	1	1 1/8	2	2 7/8	7/8	1 13/16	2.4
YSS615	2.500	2.83	15	2	1	1 1/4	2	2 7/8	7/8	2	2.8
YSS616	2.667	3.00	16	2	1	1 3/8	2	2 7/8	7/8	2 5/32	3.2
YSS618	3.000	3.33	18	2	1	1 5/8	2	2 7/8	7/8	2 1/2	4.4
YSS621	3.500	3.83	21	2	1	2	2	2 7/8	7/8	3	6.5
YSS624	4.000	4.33	24	2	1 1/8	2	2	3 1/2	1 1/2	3	9.1
YSS627	4.500	4.83	27	2	1 1/8	2	2	3 1/2	1 1/2	3	10.8
YSS630	5.000	5.33	30	2	1 1/8	2 1/8	2	3 1/2	1 1/2	3 1/8	13.3
YSS633	5.500	5.83	33	2	1 1/8	2 1/4	2	3 1/2	1 1/2	3 1/4	15.8
YSS636	6.000	6.33	36	2	1 1/8	2 5/16	2	3 1/2	1 1/2	3 3/8	19.4
YSS642	7.000	7.33	42	2	1 1/4	2 5/16	2	3 1/2	1 1/2	3 3/8	25.2
YCS648	8.000	8.33	48	4	1 1/4	2 1/8	2	3 1/2	1 1/2	3 3/8	18.1
YCS654 ■	9.000	9.33	54	4	1 1/4	2 1/8	2	3 1/2	1 1/2	3 3/8	19.8
YCS660	10.000	10.33	60	4	1 1/4	2 1/4	2	3 1/2	1 1/2	3 3/4	24.0
YCS666 ■	11.000	11.33	66	4	1 1/4	2 1/4	2	3 1/2	1 1/2	3 3/4	26.1
YCS672	12.000	12.33	72	4	1 1/4	2 1/4	2	3 1/2	1 1/2	3 3/4	29.3
YCS684	14.000	14.33	84	4	1 1/4	2 1/4	2	3 1/2	1 1/2	3 3/4	35.9
YCS696	16.000	16.33	96	4	1 1/4	2 1/4	2	3 1/2	1 1/2	3 3/4	40.0
YCS6108 ■	18.000	18.33	108	4	1 1/4	2 1/4	2	3 1/2	1 1/2	3 3/4	45.9
YCS6120	20.000	20.33	120	4	1 3/8	2 3/4	2	3 1/2	1 1/2	4 1/2	56.0
YCS6132 ■	22.000	22.33	132	4	1 3/8	2 3/4	2	3 3/4	1 3/4	4 1/2	60.0
YCS6144 ■	24.000	24.33	144	4	1 3/8	2 3/4	2	3 3/4	1 3/4	4 1/2	65.0

All 42 Tooth Gears and Smaller are Steel. All Larger Sizes, Cast Iron.
■Will not be restocked; available in production quantities only when present supply is exhausted.

Table No. 2 **Stock Steel and Cast Iron Spur Gears with Split Taper Bushings**

Part No.		Diameter		No. Teeth	Type	Dimensions									Wt. Lbs. Less Bush.
Gear	Bush.	Pitch	Nominal O.D.			F	O.L.	L	P	C	H	G	X	E	
YSS6P24	P1	4.000"	4.33"	24	11	2"	2 7/8"	1 15/16"	5/8"	11/16"	3"	—	—	1/4"	5.2
YSS6P27 ■	P1	4.500	4.83	27	11	2	2 7/8	1 15/16	5/8	11/16	3	—	—	1/4	7.1
YSS6P30	P1	5.000	5.33	30	12	2	2 3/16	1 15/16	1/16U	—	3	5/8"	1 5/16"	1/4	7.7
YSS6P33	P1	5.500	5.83	33	12	2	2 3/16	1 15/16	1/16U	—	3	5/8	1 5/16	1/4	9.3
YSS6P36	P1	6.000	6.33	36	12	2	2 3/16	1 15/16	1/16U	—	3	5/8	1 5/16	1/4	11.2
YSS6P42	P1	7.000	7.33	42	12	2	2 3/16	1 15/16	1/16U	—	3	5/8	1 5/16	1/4	15.5
YCS6Q48	Q1	8.000	8.33	48	17	2	2 25/32	2 1/2	1/2	—	4 1/8	3/4	1 3/4	9/32	15.9
YCS6Q54	Q1	9.000	9.33	54	17	2	2 25/32	2 1/2	1/2	—	4 1/8	3/4	1 3/4	9/32	18.7
YCS6Q60	Q1	10.000	10.33	60	17	2	2 25/32	2 1/2	1/2	—	4 1/8	3/4	1 3/4	9/32	18.8
YCS6Q66	Q1	11.000	11.33	66	17	2	2 25/32	2 1/2	1/2	—	4 1/8	3/4	1 3/4	9/32	22.9
YCS6Q72	Q1	12.000	12.33	72	17	2	2 25/32	2 1/2	1/2	—	4 1/8	3/4	1 3/4	9/32	23.8
YCS6Q84 ■	Q1	14.000	14.33	84	17	2	2 25/32	2 1/2	1/2	—	4 1/8	3/4	1 3/4	9/32	29.5
YCS6Q96	Q1	16.000	16.33	96	17	2	2 25/32	2 1/2	1/2	—	4 1/8	3/4	1 3/4	9/32	35.0
YCS6Q108	Q1	18.000	18.33	108	17	2	2 25/32	2 1/2	1/2	—	4 1/8	3/4	1 3/4	9/32	40.0
YCS6Q120	Q1	20.000	20.33	120	17	2	2 25/32	2 1/2	1/2	—	4 1/8	3/4	1 3/4	9/32	52.5
YCS6Q132 ■	Q1	22.000	22.33	132	17	2	2 25/32	2 1/2	1/2	—	4 1/8	3/4	1 3/4	9/32	50.5
YCS6Q144 ■	Q1	24.000	24.33	144	17	2	2 25/32	2 1/2	1/2	—	4 1/8	3/4	1 3/4	9/32	55.5

All 42 Tooth Gears and Smaller are Steel. All Larger Sizes, Cast Iron.
NOTE—"U" after "P" dimension indicates bushing flange under rim by amount shown.
■Will not be restocked; available in production quantities only when present supply is exhausted.

FIGURE 12–6. Spur gears: 6 pitch, 2-in. face, 20° pressure angle. (Courtesy Browning Manufacturing Division, Emerson Electric Co.)

We can now list the gear combinations from the catalog that give 16 when their pitch diameters are added together. The velocity ratios for these combinations are also calculated.

Available Pitch Diameters with Sum of 16	$\frac{D_p}{D_g}$
4, 12	0.333
6, 10	0.6
7, 9	0.778
8, 8	1.0

Table No. 1 — **Stock Steel and Cast Iron Minimum Bore Spur Gears**

Part No.	Diameter Pitch	Diameter Nominal O.D.	No. Teeth	Type	Bore Stock	Bore Max.	F	L	P	H	Wt. Lbs.
YSS412	3.000"	3.50"	12	2	1 1/16"	1 3/8"	3 1/2"	4 1/2"	1"	2 1/4"	6.6
YSS414	3.500	4.00	14	2	1 1/16	1 7/8	3 1/2	4 1/2	1	2 3/4	9.8
YSS415	3.750	4.25	15	2	1 1/16	2	3 1/2	4 1/2	1	3	11.5
YSS416	4.000	4.50	16	2	1 15/16	2 1/4	3 1/2	4 1/2	1	3 1/4	12.7
YSS418	4.500	5.00	18	2	1 15/16	2 3/8	3 1/2	4 1/2	1	3 3/4	16.8
YSS420	5.000	5.50	20	2	1 15/16	3	3 1/2	4 1/2	1	4 1/4	21.4
YSS424	6.000	6.50	24	2	1 15/16	3 1/2	3 1/2	4 1/2	1	5 1/4	32.2
YSS428	7.000	7.50	28	2	1 15/16	4 1/2	3 1/2	4 1/2	1	6 1/4	45.3
YSS432	8.000	8.50	32	3	1 7/16	2 3/4	3 1/2	4 3/4	1 1/4	4 1/2	47.3
YCS436	9.000	9.50	36	3	1 7/16	2 3/4	3 1/2	4 3/4	1 1/4	4 1/2	52.5
YCS440	10.000	10.50	40	3	1 7/16	3 1/4	3 1/2	4 3/4	1 1/4	5 1/4	59.9
YCS444■	11.000	11.50	44	3	1 7/16	3 1/4	3 1/2	4 3/4	1 1/4	5 1/4	69.8
YCS448	12.000	12.50	48	3	1 7/16	3 1/4	3 1/2	5 1/4	1 3/4	5 1/4	78.5
YCS456	14.000	14.50	56	4	1 7/16	3 1/4	3 1/2	5 1/4	1 3/4	5 1/4	84.8
YCS460■	15.000	15.50	60	4	1 9/16	3 1/4	3 1/2	5 1/4	1 3/4	5 1/4	91.0
YCS464	16.000	16.50	64	4	1 9/16	3 1/4	3 1/2	5 1/4	1 3/4	5 1/4	93.8
YCS472	18.000	18.50	72	4	1 9/16	3 1/4	3 1/2	5 1/4	1 3/4	5 1/4	107
YCS480	20.000	20.50	80	4	1 9/16	3 1/4	3 1/2	5 1/4	1 3/4	5 1/4	114
YCS496■	24.000	24.50	96	4	1 11/16	3 3/8	3 1/2	5 1/4	1 3/4	5 3/4	157
YCS4112■	28.000	28.50	112	4	1 11/16	3 3/8	3 1/2	5 1/4	1 3/4	5 3/4	186
YCS4128■	32.000	32.50	128	4	1 15/16	3 3/8	3 1/2	5 1/4	1 3/4	5 3/4	201
YCS4160■	40.000	40.50	160	4	1 15/16	3 3/8	3 1/2	5 1/4	1 3/4	6	253

All 28 Tooth Gears and Smaller are Steel. All Larger Sizes, Cast Iron.
■Will not be restocked; available in production quantities only when present supply is exhausted.

Table No. 2 — **Stock Steel and Cast Iron Spur Gears with Split Taper Bushings**

Gear	Bush.	Diameter Pitch	Diameter Nominal O.D.	No. Teeth	Type	F	O.L.	L	P	C	H	G	X	E	Wt. Lbs. Less Bush.
YSS4Q20	Q2	5.000"	5.50"	20	11	3 1/2"	4 17/32"	3 1/2"	3/4"	3/4"	4 1/8"	—	—	9/32"	12.9
YSS4Q24	Q2	6.000	6.50	24	11	3 1/2	4 17/32	3 1/2	3/4	3/4	4 1/8	—	—	9/32	21.5
YSS4Q28	Q2	7.000	7.50	28	12	3 1/2	3 25/32	3 1/2	0	—	4 1/8	3/4"	2 3/4"	9/32	29.4
YSS4R32	R1	8.000	8.50	32	12	3 1/2	3 1/2	2 7/8	5/8*	—	5 3/8	7/8	2	9/32	28.5
YCS4R36■	R1	9.000	9.50	36	18	3 1/2	3 1/2	2 7/8	5/8	—	5 3/8	7/8	2	9/32	36.4
YCS4R40	R1	10.000	10.50	40	18	3 1/2	3 1/2	2 7/8	5/8	—	5 3/8	7/8	2	9/32	42.7
YCS4R44■	R1	11.000	11.50	44	18	3 1/2	3 1/2	2 7/8	5/8	—	5 3/8	7/8	2	9/32	47.8
YCS4R48■	R1	12.000	12.50	48	18	3 1/2	3 1/2	2 7/8	5/8	—	5 3/8	7/8	2	9/32	56.5
YCS4R56■	R1	14.000	14.50	56	18	3 1/2	3 1/2	2 7/8	5/8	—	5 3/8	7/8	2	9/32	64.0
YCS4R60■	R1	15.000	15.50	60	18	3 1/2	3 1/2	2 7/8	5/8	—	5 3/8	7/8	2	9/32	67.5
YCS4R64	R1	16.000	16.50	64	18	3 1/2	3 1/2	2 7/8	5/8	—	5 3/8	7/8	2	9/32	73.8
YCS4R72	R1	18.000	18.50	72	18	3 1/2	3 1/2	2 7/8	5/8	—	5 3/8	7/8	2	9/32	84.0
YCS4R80	R1	20.000	20.50	80	18	3 1/2	3 1/2	2 7/8	5/8	—	5 3/8	7/8	2	9/32	96.5
YCS4R96■	R1	24.000	24.50	96	18	3 1/2	3 1/2	2 7/8	5/8	—	5 3/8	7/8	2	9/32	124
YCS4S112	S1	28.000	28.50	112	17	3 1/2	4 3/4	4 3/8	7/8	—	6 3/8	1 1/16	3 5/16	3/8	166
YCS4S128	S1	32.000	32.50	128	17	3 1/2	4 3/4	4 3/8	7/8	—	6 3/8	1 1/16	3 5/16	3/8	199
YCS4S160	S1	40.000	40.50	160	17	3 1/2	4 3/4	4 3/8	7/8	—	6 3/8	1 1/16	3 5/16	3/8	250

All 28 Tooth Gears and Smaller are Steel. All Larger Sizes, Cast Iron. *Face of bushing flange is under rim by amount shown.
■Will not be restocked; available in production quantities only when present supply is exhausted.

FIGURE 12–7. Spur gears: 4 pitch, 3¼-in. face, 20° pressure angle. (Courtesy Browning Manufacturing Division, Emerson Electric Co.)

The gear combination that comes closest to meeting the desired velocity ratio of 0.708 has a 7-in.-pitch-diameter pinion and a 9-in.-pitch-diameter gear.

The selection of the exact gear from the catalog would depend on other factors, such as the mounting method used. The manufacturer provides two types of gears with these pitch diameters, as shown in Tables 1 and 2 of Figure 12–3.

Of interest in these catalog pages is the tooth form contours shown in the upper-right corner of each page. While not to scale, they do show the *relative* tooth size differences existing in the various diametral pitches. Where gear tooth loads are high, a lower diametral pitch and larger teeth would be selected.

▶ **PROBLEMS**

12–1 Name the three standard pressure angles used for spur gears.

12–2 Which of these three pressure angles is obsolescent?

12–3 Calculate the addendum of a 16 diametral pitch, 20° pressure angle spur gear.

12–4 Calculate the outside diameter of the gear in Problem 12–3 using a pitch diameter of 2 in. Compare the calculated outside diameter with the nominal OD listed in the catalog page in Figure 12–3 for the same sized gear.

12–5 A 6 diametral pitch, 2-in.-pitch-diameter spur gear meshes with a 6-in.-pitch-diameter spur gear. The pressure angle is 20°. Make a graphical layout to determine if interference occurs.

12–6 Figure 12–2 shows the layout made to check interference for two gears having pitch diameters of $3\frac{1}{2}$ and 4 in. Change the pinion pitch diameter to 2 in. and check for interference. Use the same diametral pitch, pressure angle, and gear pitch diameter as in Figure 12–2. What is the effect of reducing the pinion pitch diameter?

12–7 Calculate the addendum, dedendum, clearance, and whole depth of a $14\frac{1}{2}°$, full-depth involute gear tooth. The diametral pitch is 12.

12–8 Calculate the addendum, dedendum, clearance, and whole depth of a 20° involute gear tooth having a diametral pitch of 12. Compare the results with the values found in Problem 12–7.

12–9 Using the catalog data, select two spur gears that can be mounted on shafts that are exactly $1\frac{1}{4}$ in. apart. The velocity ratio is unimportant.

12–10 The pinion of two mating spur gears has 24 teeth and a 20° pressure angle. The center distance is $3\frac{1}{2}$ in. Using the catalog data, select the mating gear and determine the diametral pitch. (*Hint:* Use Equation 11–6 and try various combinations.)

12–11 Select two gears of 8 diametral pitch with a velocity ratio as close as possible to 0.7857. Use the catalog data.

12–12 A Browning part no. YSS836 spur gear is to be used in a gear box. Find the addendum, dedendum, clearance, and whole depth of the tooth form.

CHAPTER 13
HELICAL AND HERRINGBONE GEARS

13-1 USE

Helical gears are inherently quieter in operation than spur gears, and for this reason are used in preference to spur gears when the shafts are parallel. For a given noise level, this means that the helical gear can be operated at a higher angular velocity than the spur gear. The sliding tooth action that occurs when helical gears mesh makes the helical gear more tolerant of shock loading and gives it more load-carrying capacity than a spur gear of the same size. A disadvantage of the helical gear is the *axial*, or thrust, load that occurs.

Table 11-1 classified helical gears as being used for both parallel and *skew* shaft relationships. Thus, we may say that another use of helical gears is for the transmission of motion between skew shafting. Figure 13-1 shows two helical gears that would be mounted on skew shafts. The reader is referred to Figure 11-9 for the application of helical gearing to parallel shafts.

The uses for helical gearing can be summarized as follows.

1. **Helical gears are used in applications where quiet operation is important.**
2. **Helical gears are preferred over spur gears when loads and speeds are high.**
3. **Helical gears are used for both parallel and skew shafts.**

13-2 AXIAL LOAD OF HELICAL GEARS

The tooth form of the helical gear geometrically is a helix. It can be considered in the same manner as the helix of the screw thread. This geometrical identity is shown in Figure 13-2. Here a schematic representation of a right-hand helical gear is compared with a conventional drawing of a right-hand screw thread. The *helix* angle,

FIGURE 13–1. Helical gears for skew shaft mounting. (Courtesy Browning Manufacturing Division, Emerson Electric Co.)

ψ (psi), is shown on the helical gear and the corresponding angle identified on the screw thread. Figure 13–2 explains how to identify the *hand* of the gear. A helical gear is either right- or left-hand, as is the screw thread. To identify the hand of a helical gear, it is important to remember that it must be viewed along the gear axis.

Tooth contact along the helix angle results in an axial load being transmitted from the gears to the shafts when the gears are mounted on parallel shafts. This condition is shown in Figure 13–3. The direction of rotation of the gears is shown by the vertical arrows located on each gear. In Figure 13–3(a), the right-hand driver exerts an axial load (thrust) to the right on the shaft; the load exerted by the driven gear is in the opposite direction.

The direction of the thrust loads can be determined by using the right-hand or left-hand rule. This rule uses the fingers and thumb of one hand to indicate the direction of rotation and the direction of the thrust, respectively. It follows here.

FIGURE 13–2. Comparison of right-hand helical gear (a) and right-hand screw (b).

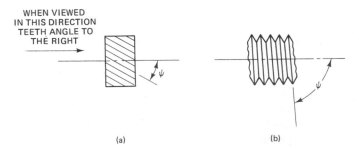

(a)   (b)

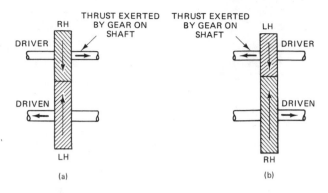

FIGURE 13–3. Axial loading of helical gears.

1. **Select the driver of the gear pair.**
2. **Use the same hand as the hand of the gear. In Figure 13–3(a), the driver is right-hand so the right hand is used.**
3. **Turn the hand so that the fingers point in the direction of rotation of the gear.**
4. **In this position the thumb then points in the direction of the thrust. It is important to remember that this load is exerted *by* the gear on the shaft.**
5. **The thrust load for the driven gear is always in the opposite direction from that of the driver.**

Figure 13–3(b) shows the thrust loads resulting when the driver is left-hand.

The thrust load existing with helical gearing is a disadvantage, since it means that the shaft bearings have to be designed to resist the thrust loads. The *herringbone gear* is designed to eliminate axial loading. Its design principle uses essentially two helical gears of opposite hand, which cancel out the axial loads in each direction. Figure 13–4 illustrates a herringbone gear.

13–3 NOMENCLATURE AND DESIGN

The terms and most of the mathematical relationships brought out in Chapter 11 apply to helical gears. The introduction of the helix angle ψ requires that consideration be given to tooth properties normal to the helix angle. This is illustrated by Figure 13–5, which introduces the *normal* circular pitch of the helical gear and shows

FIGRE 13–4.
Herringbone gear.

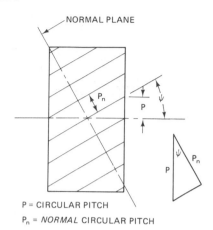

P = CIRCULAR PITCH
P_n = *NORMAL* CIRCULAR PITCH

FIGURE 13–5.   Helical gear nomenclature.

the relationship of circular pitch to normal circular pitch. The circular pitch for a helical gear has the same significance as that for a spur gear; it is the distance measured along the pitch circle from a point on a tooth to a corresponding point on an adjacent tooth in the *plane of rotation*. The *normal* circular pitch, however, is measured in a plane normal to the tooth.

Using p_n to represent normal circular pitch, the triangle in Figure 13–5 can be used to write an equation relating p and p_n:

$$p_n = p \cos \psi \qquad (13\text{--}1)$$

From Chapter 11 these terms and equations apply in exactly the same manner as for spur gears.

EXAMPLE
Diametral pitch, P, is expressed as follows:

$$P = \frac{n}{D} \qquad (11\text{--}1)$$

where n = number of teeth

D = pitch diameter

The circular pitch, p, when multiplied by the diametral pitch, P, is equal to π. Or

$$pP = \pi \qquad (11\text{--}3)$$

The center distance, L_c, is described by the following

equation:

$$L_c = \frac{D_p}{2} + \frac{D_g}{2} \qquad (11\text{--}4)$$

where D_p = pitch diameter of the pinion

D_g = pitch diameter of the gear

The product of the *normal* diametral pitch and the *normal* circular pitch equals π. The equation is

$$p_n P_n = \pi \qquad (13\text{--}2)$$

Equations 13–1, 13–2, and 13–3 can be used to obtain a fourth equation relating the normal diametral pitch and the diametral pitch. From Equation 13–2,

$$p_n P_n = \pi$$

Then

$$P_n = \frac{\pi}{p_n} \qquad (13\text{--}3)$$

From Equation 13–1,

$$p_n = p \cos \psi$$

Substituting for p_n in Equation 13–3,

$$p_n = \frac{\pi}{p \cos \psi} \qquad (13\text{--}4)$$

But $p = \pi/P$ also, and this term is substituted in Equation 13–4, giving

$$P_n = \frac{\pi}{\pi/P(\cos \psi)}$$

Simplifying, $P = P_n \cos \psi \qquad (13\text{--}5)$

Another equation for the center distance can be had by using Equations 13–5 and 11–6. Equation 11–6 is

$$L_c = \frac{n_p}{2P} + \frac{n_g}{2P} \qquad (11\text{--}6)$$

Substituting $P_n \cos \psi$ for P, we have

$$L_c = \frac{n_p}{2P_n \cos \psi} + \frac{n_g}{2P_n \cos \psi} \qquad (13\text{--}6)$$

The *angular velocity ratio* of two mating helical gears is equal to the *inverse* ratio of the number of teeth on each gear, as in spur gearing. Equation 11–11, repeated here, states this mathematically.

$$\frac{\omega_g}{\omega_p} = \frac{n_p}{n_g} \qquad (11-11)$$

Expressing the angular velocity ratio in terms of the pitch diameters requires consideration of the helix angle. From Equations 11–1 and 13–5,

$$n = PD = P_n \cos \psi D$$

When this term is substituted in Equation 11–11 for *n*, the equation becomes

$$\frac{\omega_g}{\omega_p} = \frac{D_p \cos \psi_p}{D_g \cos \psi_g} \qquad (13-7)$$

Only when the helix angles of the two gears are the same is the velocity ratio inversely proportional to the pitch diameters.

The shaft relationship, whether parallel or skew, affects the design and selection of helical gears. Figure 13–3 shows helical gears mounted on parallel shafts. In each case a right-hand gear mates with a left-hand gear, and the helix angles of the mating gears are equal. From this we can state the following requirement for parallel shafts.

Two helical gears connecting parallel shafts must have the same helix angle and must be of *opposite* hand.

When two helical gears connect skew shafts, they may or may not be of the same hand, and the helix angles are usually not the same. Consider the two left-hand helical gears shown in Figure 13–6(a). The shaft angle θ is the sum of the helix angles. In Figure 13–6(b), a right-hand pinion mates with a left-hand gear; for this condition the shaft angle is the difference of the helix angles. We can summarize as follows:

1. **When two helical gears of the same hand connect skew shafts, the shaft angle is the *sum* of the helix angles.**
2. **When two helical gears of the *opposite* hand connect skew shafts, the shaft angle is the *difference* of the helix angles.**

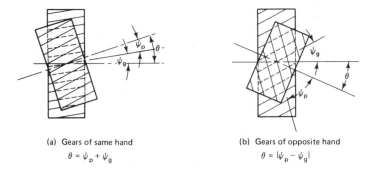

(a) Gears of same hand

$\theta = \psi_p + \psi_g$

(b) Gears of opposite hand

$\theta = |\psi_p - \psi_g|$

FIGURE 13–6.   Shaft relationship of helical gears.

The *normal pressure angle*, ϕ_n, of the helical gear is measured in a plane normal to the tooth. It corresponds to the pressure angle, ϕ, of the spur gear. The minimum number of teeth required to prevent interference on parallel helical gears depends on the normal pressure angle and also the helix angle, as shown from the data in Table 13–1.

TABLE 13–1.   Minimum Number of Teeth on a Helical Pinion to Avoid Interference

Helix Angle (degrees)	Normal Pressure Angle, ϕ_n		
	$14\frac{1}{4}°$	20°	25°
5	32	24	15
10	31	23	14
15	29	22	14
20	27	20	13
30	22	16	10
45	12	9	6

Data extracted from AGMA 207.06 and reprinted courtesy of American Gear Manufacturers Association.

Use of the principles discussed in this section is illustrated in the following examples.

EXAMPLE
It is desired to reduce the speed of an electric motor from 1800 to 750 rpm by using a pair of helical gears. Determine

suitable gear sizes, specifying the number of teeth, pitch diameters, diametral pitch pressure angle, and the helix angles.

Solution

Gear reduction of this general type would normally use parallel shafting rather than skew. The solution then will be based on parallel shafts; thus, one right-hand and one left-hand gear must be used, and the helix angles of both gears must be the same. The selection of the helix angle and diametral pitch should be such that the number of teeth will be large enough to prevent interference. Applying Equation 11–11 indirectly through use of the velocity ratio ω_g/ω_p, we can substitute the desired speeds to give

$$\frac{\omega_g}{\omega_p} = \frac{750}{1800} = \frac{n_p}{n_g}$$

A common divisor of $\frac{750}{1800}$ is 5. In addition, some multiples of 5 provide whole numbers in the numerator and denominator of the fraction. These can be used for the number of teeth of the pinion and gear. For example, using 50 as a divisor, the fraction reduces to $\frac{15}{36}$. A possible solution then is 15 teeth for the pinion and 36 teeth for the gear.

Referring to Table 13–1, the following combinations are suitable for a 15-tooth pinion.

20° helix angle and 25° pressure angle
30° helix angle and 25° pressure angle
45° helix angle with $14\frac{1}{2}$, 20,
and 25° pressure angle

Selection of a 30° helix angle and a 25° pressure angle provides low thrust loading (a lower helix angle lessens the thrust load) and adequate assurance against interference.

A diametral pitch of 8 is selected and the pitch diameters computed. For the pinion,

$$D_p = \frac{n_p}{P} = \frac{15}{8} = 1\tfrac{7}{8} \text{ in.}$$

For the gear,

$$D_g = \frac{36}{8} = 4\tfrac{1}{2} \text{ in.}$$

The final gear specifications are as follows:

> Diametral pitch: 8
> Pinion number of teeth: 15
> Gear number of teeth: 36
> 30° helix angle
> 25° pressure angle
> Pinion pitch diameter $1\frac{7}{8}$ in.
> Gear pitch diameter: $4\frac{1}{2}$ in.

It is emphasized that this is only one combination that would work in this application. Numerous other combinations would work just as well.

EXAMPLE

Two skew shafts are at an angle of 10° with each other. A right-hand helical gear of 8 diametral pitch is keyed to one shaft and has 24 teeth. The second shaft is to be driven at one half the speed of the first shaft. Assuming that the 24-tooth pinion on the first shaft has a 45° helix angle, what is the hand of the mating gear? Also find (a) its helix angle, (b) the number of teeth on the gear, and (c) the shaft center distance.

Solution

The 10° shaft angle is so small that it must be obtained using the difference of two helix angles, thus requiring opposite hand gears. The mating gear therefore has to be a left-hand gear. Since it is desirable to keep a lower helix angle, the helix angle of the gear would be 45° − 10°, or 35°.

Applying Equation 11–11, with an angular velocity ratio of $\frac{1}{2}$, the number of teeth on the gear is $2 \times 24 = 48$.

Equation 11–6 can be used to determine the center distance. It is

$$L_c = \frac{n_p}{2P} + \frac{n_g}{2P}$$

$$= \frac{24}{2(8)} + \frac{48}{2(8)}$$

$$= 4\frac{1}{2} \text{ in.}$$

13–4 SELECTION OF HELICAL GEARS

Standard helical gears are also available as catalog items from various manufacturers. However, because helical gears are frequently used as matched pairs designed for a specific application, they are more likely to be fabricated especially for the application.

Figure 13–7 shows a catalog page of one gear manufacturer, the Browning Manufacturing Division of the Emerson Electric Co. It lists the availability of 45° helix angle left- and right-hand $14\frac{1}{2}°$ pressure angle helical gears, all of 6 diametral pitch. With only a 45° helix angle, the shaft positions are restricted to parallel and right angle (90°).

FIGURE 13–7. Helical gears: 6 diametral pitch, 8.48 normal pitch, 45° helix, $14\frac{1}{2}°$ pressure angle. (Courtesy Browning Manufacturing Division, Emerson Electric Co.)

Table No. 1 — **Stock Hardened Steel Helical Gears**

Right Hand	Left Hand	Pitch	Outside	No. Teeth	Type	Bore	F	Keyway	Wt. Lbs.
NSH608AR ■	NSH608AL ■	1.333″	1.569″	8	H2	$\frac{5}{8}$″	$1\frac{1}{4}$″	$\frac{1}{8}$″x $\frac{1}{16}$″	.4
NSH610AR ■	NSH610AL ■	1.667	1.903	10	H2	$\frac{3}{4}$	$1\frac{1}{4}$	$\frac{3}{16}$ x $\frac{3}{32}$	.6
NSH612AR	NSH612AL ■	2.000	2.236	12	H2	1	$1\frac{1}{4}$	$\frac{1}{4}$ x $\frac{1}{8}$	.8
NSH615AR ■	NSH615AL ■	2.500	2.736	15	H2	1	$1\frac{1}{4}$	$\frac{1}{4}$ x $\frac{1}{8}$	1.4
NSH618AR ■	NSH618AL ■	3.000	3.236	18	H2	1	$1\frac{1}{4}$	$\frac{1}{4}$ x $\frac{1}{8}$	2.3
NSH624AR	NSH624AL ■	4.000	4.236	24	H2	1	$1\frac{1}{4}$	$\frac{1}{4}$ x $\frac{1}{8}$	4.2

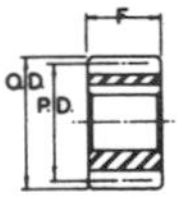

TYPE H2

Gears with 8 through 15 Teeth are hardened all over; Others have hardened teeth only.
■Will not be restocked; available in production quantities only when present supply is exhausted.

Table No. 2 — **Stock Hardened Steel Helical Gears**

Right Hand	Left Hand	Pitch	Outside	No. Teeth	Type	Bore	F	L	P	H	Keyway	Wt. Lbs.
NSH608R ■	NSH608L ■	1.333″	1.569″	8	H3	$\frac{5}{8}$″	1″	$1\frac{3}{4}$″	$\frac{3}{4}$″	1″	$\frac{1}{8}$″x $\frac{1}{16}$″	.3
NSH609R ■	NSH609L ■	1.500	1.736	9	H3	$\frac{3}{4}$	1	$1\frac{3}{4}$	$\frac{3}{4}$	$1\frac{11}{64}$	$\frac{3}{16}$ x $\frac{3}{32}$	.5
NSH610R ■	NSH610L ■	1.667	1.903	10	H3	$\frac{3}{4}$	1	$1\frac{3}{4}$	$\frac{3}{4}$	$1\frac{11}{32}$	$\frac{3}{16}$ x $\frac{3}{32}$	.6
NSH612R	NSH612L	2.000	2.236	12	H3	1	1	$1\frac{3}{4}$	$\frac{3}{4}$	$1\frac{5}{8}$	$\frac{1}{4}$ x $\frac{1}{8}$	.9
NSH615R	NSH615L	2.500	2.736	15	H3	$1\frac{1}{4}$	1	$1\frac{3}{4}$	$\frac{3}{4}$	2	$\frac{5}{16}$ x $\frac{5}{32}$	1.4
NSH618R	NSH618L	3.000	3.236	18	H3	$1\frac{1}{4}$	1	$1\frac{3}{4}$	$\frac{3}{4}$	$2\frac{1}{4}$	$\frac{5}{16}$ x $\frac{5}{32}$	2.3
NSH620R ■	NSH620L ■	3.333	3.569	20	H3	$1\frac{1}{4}$	1	$1\frac{3}{4}$	$\frac{3}{4}$	$2\frac{1}{2}$	$\frac{5}{16}$ x $\frac{5}{32}$	2.9
NSH624R	NSH624L	4.000	4.236	24	H3	$1\frac{1}{4}$	1	$1\frac{3}{4}$	$\frac{3}{4}$	$2\frac{1}{2}$	$\frac{5}{16}$ x $\frac{5}{32}$	3.9
NSH630R	NSH630L	5.000	5.236	30	H4	$1\frac{1}{4}$	1	$1\frac{3}{4}$	$\frac{3}{4}$	$2\frac{1}{2}$	$\frac{5}{16}$ x $\frac{5}{32}$	5.1
NSH636R	NSH636L	6.000	6.236	36	H4	$1\frac{1}{4}$	1	$1\frac{3}{4}$	$\frac{3}{4}$	$2\frac{1}{2}$	$\frac{5}{16}$ x $\frac{5}{32}$	6.8

All of above Gears have Setscrew at 90° to Keyway.
Gears with 8 through 15 Teeth are hardened all over; Others have hardened teeth only.
■Will not be restocked; available in production quantities only when present supply is exhausted.

Table No. 3 — **Stock Bronze Helical Gears**

Right Hand	Left Hand	Pitch	Outside	No. Teeth	Type	Bore	F	L	P	H	Keyway	Wt. Lbs.
NBH608R ■	NBH608L ■	1.333″	1.569″	8	H3	$\frac{5}{8}$″	1″	$1\frac{3}{4}$″	$\frac{3}{4}$″	1″	$\frac{1}{8}$″x $\frac{1}{16}$″	.4
NBH609R ■	NBH609L ■	1.500	1.736	9	H3	$\frac{3}{4}$	1	$1\frac{3}{4}$	$\frac{3}{4}$	$1\frac{11}{64}$	$\frac{3}{16}$ x $\frac{3}{32}$	.5
NBH610R ■	NBH610L ■	1.667	1.903	10	H3	1	1	$1\frac{3}{4}$	$\frac{3}{4}$	$1\frac{11}{32}$	$\frac{3}{16}$ x $\frac{3}{32}$	.8
NBH612R ■	NBH612L ■	2.000	2.236	12	H3	1	1	$1\frac{3}{4}$	$\frac{3}{4}$	$1\frac{5}{8}$	$\frac{1}{4}$ x $\frac{1}{8}$	.9
NBH615R	NBH615L	2.500	2.736	15	H3	$1\frac{1}{4}$	1	$1\frac{3}{4}$	$\frac{3}{4}$	2	$\frac{5}{16}$ x $\frac{5}{32}$	1.5
NBH618R	NBH618L	3.000	3.236	18	H3	$1\frac{1}{4}$	1	$1\frac{3}{4}$	$\frac{3}{4}$	$2\frac{1}{4}$	$\frac{5}{16}$ x $\frac{5}{32}$	2.3
NBH620R ■	NBH620L ■	3.333	3.569	20	H3	$1\frac{1}{4}$	1	$1\frac{3}{4}$	$\frac{3}{4}$	$2\frac{1}{2}$	$\frac{5}{16}$ x $\frac{5}{32}$	3.0
NBH624R ■	NBH624L ■	4.000	4.236	24	H4	$1\frac{1}{4}$	1	$1\frac{3}{4}$	$\frac{3}{4}$	$2\frac{1}{2}$	$\frac{5}{16}$ x $\frac{5}{32}$	3.8
NBH630R	NBH630L	5.000	5.236	30	H4	$1\frac{1}{4}$	1	$1\frac{3}{4}$	$\frac{3}{4}$	$2\frac{1}{2}$	$\frac{5}{16}$ x $\frac{5}{32}$	5.1
NBH636R	NBH636L	6.000	6.236	36	H4	$1\frac{1}{4}$	1	$1\frac{3}{4}$	$\frac{3}{4}$	$2\frac{1}{2}$	$\frac{5}{16}$ x $\frac{5}{32}$	6.5

All of above Gears have Setscrew at 90° to Keyway.
■Will not be restocked; available in production quantities only when present supply is exhausted.

10 DIAMETRAL PITCH
14.14 NORMAL PITCH

45° HELIX
14½° PRESSURE ANGLE

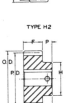

TYPE H2

TYPE H3

TYPE H4

Table No. 1 — Stock Hardened Steel Helical Gears

Right Hand	Left Hand	Pitch	Outside	No. Teeth	Type	Bore	F	Keyway	Wt. Lbs.
NSH1008AR ■	NSH1008AL ■	.800"	.941"	8	H2	3/8"	7/8"	1/16"x 1/32"	.1
NSH1010AR	NSH1010AL	1.000	1.141	10	H2	1/2	7/8	1/8 x 1/16	.1
NSH1012AR	NSH1012AL	1.200	1.341	12	H2	5/8	7/8	1/8 x 1/16	.2
NSH1015AR	NSH1015AL	1.500	1.641	15	H2	3/4	7/8	3/16 x 3/32	.3
NSH1020AR	NSH1020AL	2.000	2.141	20	H2	3/4	7/8	3/16 x 3/32	.6
NSH1025AR	NSH1025AL	2.500	2.641	25	H2	3/4	7/8	3/16 x 3/32	1.0
NSH1030AR	NSH1030AL	3.000	3.141	30	H2	3/4	7/8	3/16 x 3/32	1.6
NSH1040AR	NSH1040AL	4.000	4.141	40	H2	3/4	7/8	3/16 x 3/32	3.0

Gears with 8 and 20 Teeth are hardened all over; Others have hardened teeth only.

8 DIAMETRAL PITCH
11.31 NORMAL PITCH

45° HELIX
14½° PRESSURE ANGLE

Table No. 2 — Stock Hardened Steel Helical Gears

Right Hand	Left Hand	Pitch	Outside	No. Teeth	Type	Bore	F	Keyway	Wt. Lbs.
NSH808AR ■	NSH808AL ■	1.000"	1.177"	8	H2	1/2"	1"	1/8"x 1/16"	.1
NSH810AR	NSH810AL	1.250	1.427	10	H2	5/8	1	1/8 x 1/16	.2
NSH812AR ■	NSH812AL ■	1.500	1.677	12	H2	3/4	1	3/16 x 3/32	.3
NSH816AR	NSH816AL	2.000	2.177	16	H2	7/8	1	3/16 x 3/32	.7
NSH820AR	NSH820AL	2.500	2.677	20	H2	7/8	1	3/16 x 3/32	1.2
NSH824AR	NSH824AL	3.000	3.177	24	H2	7/8	1	3/16 x 3/32	1.8
NSH832AR ■	NSH832AL ■	4.000	4.177	32	H2	7/8	1	3/16 x 3/32	3.4

Gears with 8 through 12 Teeth are hardened all over; Others have hardened teeth only.

Table No. 3 — Stock Hardened Steel Helical Gears

Right Hand	Left Hand	Pitch	Outside	No. Teeth	Type	Bore	F	L	P	H	Keyway	Wt. Lbs.
NSH808R	NSH808L	1.000"	1.177"	8	H3	1/2"	3/4"	1 1/4"	1/2"	3/4"	1/8"x 1/16"	.2
NSH810R	NSH810L	1.250	1.427	10	H3	5/8	3/4	1 1/4	1/2	1	1/8 x 1/16	.3
NSH812R	NSH812L	1.500	1.677	12	H3	3/4	3/4	1 1/4	1/2	1 1/4	3/16 x 3/32	.5
NSH816R	NSH816L	2.000	2.177	16	H3	1	3/4	1 1/4	1/2	1 5/8	1/4 x 1/8	.6
NSH820R	NSH820L	2.500	2.677	20	H3	1	3/4	1 1/4	1/2	2	1/4 x 1/8	1.1
NSH824R	NSH824L	3.000	3.177	24	H3	1	3/4	1 1/4	1/2	2	1/4 x 1/8	1.6
NSH832R ■	NSH832L ■	4.000	4.177	32	H4	1	3/4	1 1/4	1/2	2	1/4 x 1/8	2.5
NSH840R	NSH840L	5.000	5.177	40	H4	1	3/4	1 1/4	1/2	2	1/4 x 1/8	4.0
NSH848R	NSH848L	6.000	6.177	48	H4	1	3/4	1 1/4	1/2	2 1/4	1/4 x 1/8	4.6

All of above Gears have Setscrew at 90° to Keyway.
Gears with 8 through 16 Teeth are hardened all over; Others have hardened teeth only.

Table No. 4 — Stock Bronze Helical Gears

Right Hand	Left Hand	Pitch	Outside	No. Teeth	Type	Bore	F	L	P	H	Keyway	Wt. Lbs.
NBH808R	NBH808L	1.000"	1.177"	8	H3	1/2"	3/4"	1 1/4"	1/2"	3/4"	1/8"x 1/16"	.1
NBH810R	NBH810L	1.250	1.427	10	H3	5/8	3/4	1 1/4	1/2	1	1/8 x 1/16	.3
NBH812R ■	NBH812L ■	1.500	1.677	12	H3	3/4	3/4	1 1/4	1/2	1 1/4	3/16 x 3/32	.4
NBH816R	NBH816L	2.000	2.177	16	H3	1	3/4	1 1/4	1/2	1 5/8	1/4 x 1/8	.7
NBH820R ■	NBH820L ■	2.500	2.677	20	H3	1	3/4	1 1/4	1/2	2	1/4 x 1/8	1.2
NBH824R	NBH824L	3.000	3.177	24	H3	1	3/4	1 1/4	1/2	2	1/4 x 1/8	1.7
NBH832R ■	NBH832L ■	4.000	4.177	32	H4	1	3/4	1 1/4	1/2	2	1/4 x 1/8	2.4
NBH840R ■	NBH840L ■	5.000	5.177	40	H4	1	3/4	1 1/4	1/2	2	1/4 x 1/8	3.1
NBH848R ■	NBH848L ■	6.000	6.177	48	H4	1	3/4	1 1/4	1/2	2 1/4	1/4 x 1/8	4.8

All of above Gears have Setscrew at 90° to Keyway.
■Will not be restocked; available in production quantities only when present supply is exhausted.

FIGURE 13–8. Helical gears: 10 diametral pitch, 45° helix, 14½° pressure angle; and 8 diametral pitch, 45° helix, 14½° pressure angle. (Courtesy Browning Manufacturing Division, Emerson Electric Co.)

Figure 13–8 is another page from the same catalog showing the 8 and 10 diametral pitch gears available. Note that these gears are supplied only with the 45° helix angle and 14½° pressure angle, an indication of the more restricted availability of the standard helical gear as compared to the spur gear.

► PROBLEMS

13–1   Name two of the three main reasons that helical gears may be used in preference to spur gears.

13–2   Name a disadvantage of the helical gear.

13–3   The herringbone gear is sometimes used in place of the helical gear in order to _____. (Fill in the correct answer.)

13–4   Identify the hand of the two gears in Figure 13–1.

13–5   Determine the direction of the thrust load exerted by the gear on the shaft for each of the two helical gears in Figure 13–9. Rotation is as shown for the large gear.

13–6   An 8 diametral pitch helical gear has a helix angle of 35°. Determine (a) the circular pitch, (b) the normal circular pitch, and (c) the normal diametral pitch.

13–7   A helical gear with a helix angle of 35° meshes with another helical gear with a 45° helix angle. The pitch diameters of the two gears are 6 and 8 in., respectively. Determine the angular velocity ratio and the center distance.

13–8   The speed of an electric motor is to be reduced from 3600 to 900 rpm by using two helical gears. The shafting is parallel, and the center distance is to be as close to $10\frac{1}{2}$ in. as possible. Determine two gears suitable for the application, specifying the diametral pitch, number of teeth, helix angles, pitch diameters, and final center distance.

13–9   Two gears of opposite hand have the same helix angle. What must be the shaft angle if the two gears are mated?

13–10   A right-hand helical gear with a helix angle of 45° meshes with another right-hand gear with a 45° helix angle. What is the angle between shafts?

13–11   Using Figures 13–7 and 13–8, select two helical gears that have a velocity ratio of 0.8 and a center distance of $2\frac{1}{4}$ in.

13–12   Using Figures 13–7 and 13–8, select two gears with a velocity ratio of 0.833 and a center distance as close to 3 in. as possible.

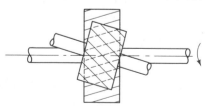

FIGURE 13–9

CHAPTER 14
BEVEL, HYPOID, AND WORM GEARS

14-1   INTRODUCTION

Although bevel, hypoid, and worm gears may be considered to be specialized gear types requiring higher technology in their design and manufacture, they are all extensively used in mechanisms. The *bevel gear*, and its variations, is the only type that can be used when shaft axes intersect. The *worm gear* provides very high speed reductions for nonintersecting shafts. The *hypoid gear* is the common pinion and ring gear combination of the automobile rear axle differential.

14-2   BEVEL GEAR

Bevel gears were briefly discussed in Chapter 11, and the straight and spiral bevel gears illustrated in Figures 11–6 and 11–7 are examples. Bevel gears kinematically can be considered equivalent to rolling cones. The terminology of bevel gears can be explained by referring to Figure 14–1, which shows the cross section of a bevel gear with the gear teeth on each side not sectioned. The *pitch angle* determines the theoretical rolling cone surface of the gear, and major dimensions use this surface, called the *pitch cone*, as a primary reference point.

It is apparent in Figure 14–1 that the *tooth cross section* varies across the tooth face, whereas in spur and helical gearing it is constant. However, the tooth form is still involute, and we can develop its proportions if the same relative point on the face width is used. This point is at the intersection of the *back cone* with the *pitch cone*. From this point we measure the *addendum*, *dedendum*, and *pitch diameter*, as shown in Figure 14–1.

The design and selection of bevel gears for a specific application involve finding the pitch angles of both gears. Consider

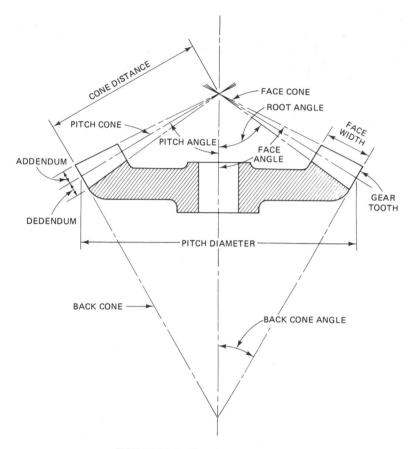

FIGURE 14–1   Bevel gear terminology.

the two external bevel gears shown in Figure 14–2. It is obvious that the shaft angle, θ, is equal to the sum of the pitch angles of the two gears. Using the identified quantities as shown in Figure 14–2, we can develop an equation relating the tangent of the pitch angle of one gear to the shaft angle and the pitch radii of the two gears. The derivation of this equation follows.

Using the distance OP as the hypotenuse of a right triangle, with r_p and r_g as opposite sides,

$$r_p = OP \sin \phi$$

and

$$r_g = OP \sin \beta$$

Dividing one equation by the other, we have

$$\frac{r_p}{r_g} = \frac{OP \sin \phi}{OP \sin \beta} = \frac{\sin \phi}{\sin \beta}$$

Solving tor sin ϕ,

$$\sin \phi = \frac{r_p}{r_g} \sin \beta \qquad (14\text{--}1)$$

But

$$\phi + \beta = \theta$$

and

$$\beta = \theta - \phi$$

Substituting this term for β in Equation 14–1 gives

$$\sin \phi = \frac{r_p}{r_g} \sin (\theta - \phi) \qquad (14\text{--}2)$$

The term sin $(\theta - \phi)$ can be changed by substitution of a trigonometric identity. The identity is

$$\sin (\theta - \phi) - \sin \theta \cos \phi - \cos \theta \sin \phi$$

After substitution, Equation 14–2 becomes

$$\sin \phi = \frac{r_p}{r_g} (\sin \theta \cos \phi - \cos \theta \sin \phi)$$

Both sides of the equation are now divided by cos ϕ. Since sin $\phi / \cos \phi = \tan \phi$, the equation can be reduced to

$$\tan \phi = \frac{r_p}{r_g} (\sin \theta - \cos \theta \tan \phi)$$

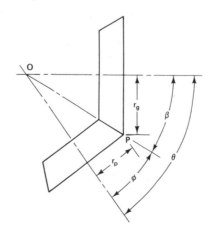

FIGURE 14–2. External bevel gear contact.

Solving for tan ϕ,

$$\tan \phi = \frac{\sin \theta}{r_g/r_p + \cos \theta} \qquad (14\text{--}3)$$

The velocity ratio of bevel gears is inversely proportional to the ratios of the pitch diameters and also the number of teeth on each gear. We can also use the pitch radius for the inverse proportion; thus, we have

$$\frac{\omega_g}{\omega_p} = \frac{n_p}{n_g} = \frac{r_p}{r_g}$$

Inverting the term r_p/r_g and substituting the equivalent terms in Equation 14–3 give other useful forms of the equation. They are

$$\tan \phi = \frac{\sin \theta}{n_g/n_p + \cos \theta} \qquad (14\text{--}4)$$

and

$$\tan \phi = \frac{\sin \theta}{\omega_p/\omega_g + \cos \theta} \qquad (14\text{--}5)$$

Figure 14–2 shows *external* bevel gears in contact. *Internal* bevel gears are also used. In this case the equations relating to shaft angle and pitch angles are changed. Figure 14–3 shows two bevel gears in internal contact. Two equations that apply are as follows: for the shaft angle,

$$\theta = \beta - \phi \qquad (14\text{--}6)$$

For the tan ϕ,

$$\tan \phi = \frac{\sin \theta}{\omega_p/\omega_g - \cos \theta} \qquad (14\text{--}7)$$

FIGURE 14–3.   Internal bevel gear contact.

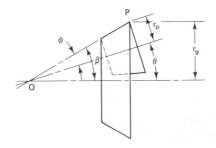

EXAMPLE

A pair of external bevel gears is to be mounted on shafts that intersect at 78°. The input speed is 1600 rpm and the output speed is 800 rpm. Determine the pitch angles of the gears.

Solution

Equation 14–5 can be used by substituting the correct values into it. After substitution, we have

$$\tan \phi = \frac{\sin 78°}{1600/800 + \cos 78°}$$

$$= \frac{0.9781}{2 + 0.2079}$$

$$= 0.4430$$

$$\phi = 23.89°$$

The pitch angle of the other gear is

$$\beta = 78 - 23.89 = 54.11°$$

14–3 SPECIAL TYPES OF BEVEL GEARS

Mitre gears are straight-tooth bevel gears that are mounted on shafts intersecting at 90° and have a 1 to 1 velocity ratio. The pitch angle of each gear is 45°, and the number of teeth on each gear is the same.

The *spiral bevel gear* shown in Figure 11–7 provides smoother action and withstands higher loads than the straight-tooth bevel gear.

A variation of the spiral bevel gear is the *hypoid gearing* shown schematically in Figure 14–4. Hypoid gearing was a major development in the automotive field, since it allowed the drive shaft to the rear axle to be lowered and provided more rear floor space in the car. At the same time, it provided quieter operation and more load-carrying ability than the straight bevel gear. An examination of Figure 14–4 shows the nonintersecting axes of the pinion and ring gear, which results in lowering the drive shaft.

Zerol bevel gears are spiral gears with a zero spiral angle. Figure 14–5 shows a Zerol bevel gear. The Zerol gear is used in place of a straight bevel gear where better performance is needed.

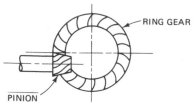

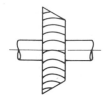

FIGURE 14–4. Hypoid gears.   FIGURE 14–5. Zerol bevel gears.

14–4 WORM GEARS

Worm gears are used to connect skew shafts that are perpendicular to each other. Figure 14–6 shows a worm reducer with its noninter-secting, perpendicular shafts. As noted previously, its advantages are the high-speed reduction that it gives and the small amount of space that it requires.

The *worm* of the worm gear reducer in Figure 14–6 is the member mounted on the lower shaft. Instead of the conventional gear teeth, the worm uses teeth that are a continuous helix. This helix is the same as that of the conventional screw thread, making the worm very similar to a screw thread. Screw thread terminology is used to describe the worm, as shown in Figure 14–7. Figure 14–7(a) shows a single-threaded worm; Figure 14–7(b) shows a double-threaded worm. Worms, like screw threads, may be threaded in multiples up to three. The double-threaded worm in Figure 14–7(b) has two threads, one opposite the other as shown in

FIGURE 14–6. Worm gear reducer. (Courtesy Winsmith Division of UMC Industries, Inc.)

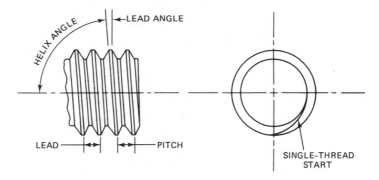

(a) Single-threaded

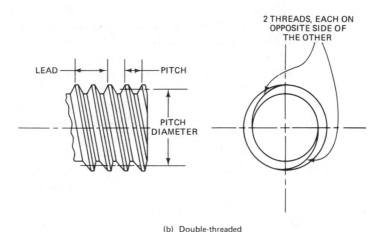

(b) Double-threaded

FIGURE 14-7. Worm terminology.

the end view. A triple-threaded worm has three threads spaced equidistant around the periphery.

Pitch of the thread is defined as the axial distance from a point on one tooth to the corresponding point on an adjacent tooth. *Lead* is defined as the distance the thread would advance in one revolution if screwed into a mating thread.

Note that for a single-threaded worm the lead is equal to the pitch; for a double-threaded worm the lead is equal to twice the pitch; and for a triple-threaded worm, the lead is equal to three times the pitch.

The angular velocity ratio of a worm gear set may be obtained by using Equation 11–11:

$$\frac{\omega_g}{\omega_p} = \frac{n_p}{n_g} \tag{11–11}$$

where ω_g = angular velocity of the gear

$\dfrac{\omega_o}{\omega_p} = 75$

ω_p = angular velocity of the worm (the worm of the worm gear set is the pinion)

n_p = number of threads on the worm

n_g = number of teeth on the gear

The number of threads refers to the single, double, or triple thread previously discussed. Thus, for a single-threaded worm, n_p is equal to 1. The low values associated with n_p are of course the reason for the large speed reductions obtained with worm gears.

In almost all cases the worm is the driver in the worm gear set. In fact, until the lead angle becomes larger than about 5°, the set is self-locking and cannot be driven by the gear.

The theoretical contact between the worm and gear is a point. Since this results in high stresses and wear, it is desirable to enlarge the contact area. This can be done by using the *throated* gears shown in Figure 14–8. This involves making the contour of the gear, or the worm and gear both, match the contour of the mating member.

Although the load-carrying ability of the worm and gear is high, the efficiency of the worm and gear is not as high as for other gearing types because of the high frictional forces developed.

▶ PROBLEMS

14–1 In a bevel gear the theoretical surface corresponding to the surface of the rolling cone is called the _____. (Fill in the correct answer.)

FIGURE 14–8. Throated worm gear sets.

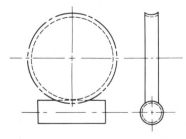

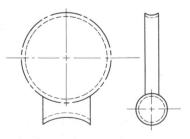

(a) Single-threaded   (b) Double-threaded

14–2   At what location are the addendum, dedendum, and pitch diameter of a bevel gear measured from?

14–3   Bevel gears that have 45° pitch angles and the same number of teeth are called _____ gears. (Fill in the correct answer.)

14–4   Why was the hypoid gear development of such importance to the automotive field?

14–5   A bevel gear having 33 teeth is the pinion of a pair of bevel gears connecting shafts at an angle of 105°. The velocity ratio is 0.75. Determine the pitch angles of the two gears and the number of teeth on the mating gear.

14–6   If the diametral pitch of the gears in Problem 14–5 is 8, find the pitch diameters of the two gears.

14–7   Two mating external bevel gears have pitch angles of 35 and 45°. Find the velocity ratio of the pair.

14–8   Two bevel gears are in internal contact as shown in Figure 14–3. If the shaft angle is 44° and the velocity ratio is 0.5, find the pitch angles of the two gears.

14–9   The worm of a worm gear set is single threaded. The gear has a diametral pitch of 8 and a pitch diameter of 6 in. What is the velocity ratio?

14–10   The gear of a worm and gear set has 64 teeth and the worm is double threaded. If the worm rotates at 2000 rpm, what is the speed of the gear?

CHAPTER 15
ORDINARY GEAR TRAINS

15–1 INTRODUCTION

Mechanism trains use various combinations of mechanisms together to perform various functions. Commonly, a mechanism train may use linkages, cams, belt drives, gearing, and other mechanisms together. A *gear train* consists of various gear combinations that have been designed into one unit to perform specified functions usually relating to output shaft speed, direction of rotation, and other factors such as torque. Figures 1–2 and 11–1 show mechanism trains in which gear trains comprise a large part of the complete mechanism. The wooden gear train of the tower clock in Figure 1–2 is used to transmit motion to the clock hands. The gear trains in the aircraft engine of Figure 11–1 drive accessory units of the engine and the magnetos used for ignition.

In the simpler gear train, all gears have their axes fixed with respect to the frame. This is called an *ordinary gear train*. In this chapter we discuss only the ordinary gear train, leaving the more complicated planetary gearing for Chapter 16.

15–2 TYPES OF ORDINARY GEAR TRAINS

The preceding section mentioned the requirement that all gear axes be fixed to the frame in order for the gear train to qualify as an ordinary gear train. Consider the worm gear motor shown in Figure 15–1. Since it has two gears in mesh, it is a gear train. The gear axes are fixed to the frame since the two shafts that constitute these axes are mounted in the frame Note that, although two gears constitute a train, in practice we are concerned with trains with larger numbers of gears.

Figure 15–2 shows the two major types of ordinary gear trains. Figure 15–2(a) shows a *simple* ordinary gear train. Figure

FIGURE 15–1. Worm gear motor. (Courtesy Winsmith Division of UMC Industries, Inc.)

15–2(b) shows a *compound* ordinary gear train. In the *simple* gear train each gear is on a separate shaft; in the *compound* gear train at least one or more shafts have two or more gears keyed to it. We should note in both cases that the gear axes are fixed with respect to the frame. Another type of compound gear train is the *reverted* gear train shown in Figure 15–3. The *reverted* gear train has at least one shaft with two or more gears keyed to it, but it also has its input and output shafts positioned so that they line up as shown in Figure 15–3. Compound gear trains in which the input and output shafts do not line up are called *nonreverted*.

FIGURE 15–2. Ordinary gear trains.

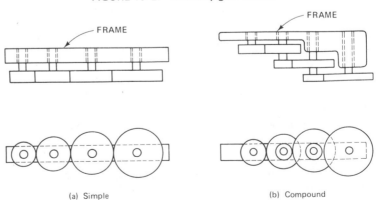

(a) Simple

(b) Compound

FIGURE 15–3. Reverted gear train. INPUT SHAFT LINES UP WITH OUTPUT

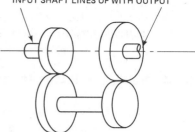

15–3 DIRECTION OF ROTATION

In the design or analysis of a gear train, it is necessary to know the direction of rotation of each gear. In an ordinary gear train in which each gear axis is fixed to the frame, it is easy to obtain the direction of rotation of each gear as long as the direction of rotation of one gear is known. Clockwise and counterclockwise (cw and ccw) are the normal descriptive terms for rotation direction and are used in this text. However, where gear trains are involved, the direction of rotation of each gear will be shown primarily by using arrows drawn on the gear itself. The method is illustrated in Figure 15–4(a), where the direction of rotation of gear *A* is shown by the arrow on gear *A*. The arrow is drawn at the point of contact pitch point with the mating gear, *B*. Since the point of contact on *B* has to have the same direction (there can be no slippage between *A* and *B*), an arrow is drawn on *B* in the same direction and adjacent to the point of contact.

To determine the direction of rotation of gear *C*, rotate the arrow on *B* until it is adjacent to the contact at *C*. Draw an arrow on *C* at the same point and in the same direction. Figure 15–4(b) shows how this method applies to a compound gear train.

An *idler* gear is used in a gear train to change the direction of rotation of the last driven gear. In addition, it may serve other purposes, such as connecting input and output shafts that are

FIGURE 15–4. Gear rotation indication.

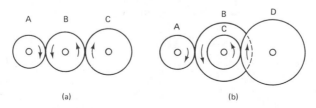

(a)   (b)

located far apart. The *idler* gear is simply a gear with the same number of teeth and the same pitch diameter as one of its mating gears. Figure 15–5 shows an example.

15–4 VELOCITY RATIO OF GEAR TRAINS

The *velocity ratio* of a gear train is the ratio of the angular velocity of the last *driven* gear to the angular velocity of the first *driving* gear. This definition is similar to that given for two mating gears, when the pinion is assumed as the driver and the gear is the driven gear. Some texts use the term *train value* in place of velocity ratio; in addition, the *reciprocal* of the velocity ratio is sometimes used in describing gear train ratios. An example is the automotive rear axle, for which terms such as "a rear axle ratio of 3.3 to 1" may be used to describe a final drive ratio for the car. The *velocity ratio* is the *reciprocal* of this, or 1/3.3. It means that the driving gear, the pinion connected to the drive shaft, turns 3.3 times for each turn of the rear axle when the car is traveling in a straight line.

The velocity ratio of a gear train is found by applying the same rules used for finding the velocity ratio of two gears. Consider the simple ordinary gear train in Figure 15–6. The gear train has three gears, *A*, *B*, and *C*, each with the number of teeth indicated. The velocity ratio of the train will be developed by working with two gears at a time, taking first gears *A* and *B*. The velocity of these two gears is

$$\frac{\omega_B}{\omega_A} = \frac{n_A}{n_B} \qquad (15\text{–}1)$$

Now consider gears *B* and *C*. The velocity ratio is

$$\frac{\omega_C}{\omega_B} = \frac{n_B}{n_C} \qquad (15\text{–}2)$$

Equations 15–1 and 15–2 are now combined by multiplying

FIGURE 15–5.   Idler gear.

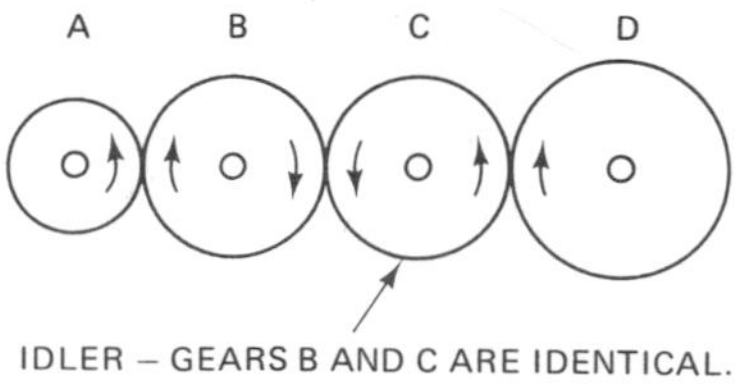

A   B   C   D

IDLER – GEARS B AND C ARE IDENTICAL.

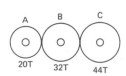

FIGURE 15–6

A   B   C

20T   32T   44T

together as follows:

$$\left(\frac{\omega_B}{\omega_A}\right)\left(\frac{\omega_C}{\omega_B}\right) = \left(\frac{n_A}{n_B}\right)\left(\frac{n_B}{n_C}\right)$$

When simplified by canceling out the terms ω_B, and n_B, the equation becomes

$$\frac{\omega_C}{\omega_A} = \frac{n_A}{n_C} \qquad\qquad (15\text{--}3)$$

Although only three gears were used to develop Equation 15–3, it can be proved that the equation holds for any number of gears as long as the train is a *simple* ordinary train. The following statement summarizes Equation 15–3.

The velocity ratio of a *simple* ordinary gear train is equal to the number of teeth of the first driving gear divided by the number of teeth of the last driven gear.

The following example illustrates an application of Equation 15–3.

EXAMPLE
The simple ordinary gear train in Figure 15–7 is driven by gear *A*. Gear *A* rotates at 1200 rpm. Find the speed of gear *D* and its direction of rotation.

Solution
Applying Equation 15–3, the velocity of the last driven gear is ω_d and

$$\frac{\omega_D}{\omega_A} = \frac{n_A}{n_D}$$

Substituting,

$$\frac{\omega_D}{1200} = \frac{24}{64}$$

$$\omega_D = (1200)\frac{24}{64}$$

$$= 450 \; rpm$$

FIGURE 15–7

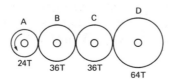

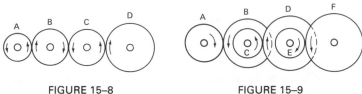

FIGURE 15–8   FIGURE 15–9

Figure 15–8 shows how the direction of rotation is determined. Gear D rotates clockwise as indicated by the arrow. Note that gears B and C have the same number of teeth, making C an idler gear.

The method developed for the simple ordinary gear train does not apply to the *compound* gear train. However, another procedure can be developed for the compound gear train. Figure 15–9 shows a compound gear train with six gears. Gear A drives gear B; C is on the same shaft as B and drives D; and E, on the same shaft as D, drives F. Using this description, the gears can be classified as *drivers* and *driven*, as in Figure 15–9.

Taking two gears at a time, we can write velocity ratio equations.

$$\frac{\omega_B}{\omega_A} = \frac{n_A}{n_B}, \qquad \frac{\omega_D}{\omega_C} = \frac{n_C}{n_D}, \qquad \frac{\omega_F}{\omega_E} = \frac{n_E}{n_F}$$

These equations can be combined by multiplying the right and left sides together, as previously. This gives the following:

$$\left(\frac{\omega_B}{\omega_A}\right)\left(\frac{\omega_D}{\omega_C}\right)\left(\frac{\omega_F}{\omega_E}\right) = \left(\frac{n_A}{n_B}\right)\left(\frac{n_C}{n_D}\right)\left(\frac{n_E}{n_F}\right) \qquad (15\text{–}4)$$

Gears B and C are on the same shaft, and D and E are on the same shaft; therefore

$$\omega_B = \omega_C \quad \text{and} \quad \omega_D = \omega_E$$

When these are substituted into the left side of Equation 15–4, it reduces to

$$\frac{\omega_F}{\omega_A} = \left(\frac{n_A}{n_B}\right)\left(\frac{n_C}{n_D}\right)\left(\frac{n_E}{n_F}\right) \qquad (15\text{–}5)$$

The term ω_F/ω_A will be recognized as the velocity ratio. In Figure 15–9, gears A, C, and E are classified as drivers, and gears B, D, and F are the driven gears. With this we can summarize Equation 15–5 as follows:

The velocity ratio of a *compound* ordinary gear train is equal to the product of the number of teeth on the drivers divided by the product of the number of teeth on the driven gears.

Equation 15–5 applies to simple ordinary gear trains as well as compound. Consider its application to the simple gear train in Figure 15–4(a). There are three gears in the train; gear A is the first driver, and gear B is the first driven gear. The last gear, C, is also a driven gear. Since C is driven by B, we must classify B as both a driver and a driven gear. Using Equation 15–5, we have

$$\frac{\omega_C}{\omega_A} = \left(\frac{n_A}{n_B}\right)\left(\frac{n_B}{n_C}\right)$$

$$= \frac{n_A}{n_C}$$

This is the same as Equation 15–3.

The situation where a gear functions as both a driver and a driven gear occurs in a compound train when an odd number of gears is used. Adding another gear to mesh with F in the train of Figure 15–9 would make F both a driver and a driven gear. When used in Equation 15–5, the number of teeth appears in both numerator and denominator, thus canceling itself out.

Example
Figures 15–10(a) shows a triple-threaded, right-hand worm driving a compound gear train. The numbers of teeth on each gear are as shown. Find the velocity ratio and the direction of rotation of gear C.

FIGURE 15–10

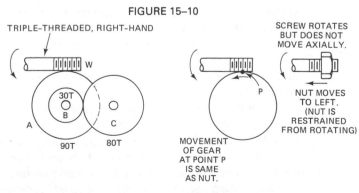

(a)

Solution

The worm is considered to be a driving gear with three teeth. Applying Equation 15–5, the velocity ratio is

$$\frac{\omega_C}{\omega_w} = \left(\frac{n_w}{n_A}\right)\left(\frac{n_B}{n_C}\right)$$

Substituting,

$$\frac{\omega_C}{\omega_w} = \left(\frac{3}{90}\right)\left(\frac{30}{80}\right)$$

$$= \frac{1}{80}$$

The direction of rotation is determined by first finding the direction of rotation of gear A. This requires consideration of the worm helix and its effect on A as the worm rotates. The screw-thread analogy in Figure 15–10(b) is helpful in analyzing this. If a right-hand screw thread rotates clockwise when viewed from the left as shown, but does not move axially, then the nut on it, which is prevented from rotating, must move to the left as shown. This is easily proved by experiment with a nut and bolt.

Point P on gear A must move as does the nut, to the left. Thus the rotation of gear A and also B is counterclockwise, while C is clockwise.

15–5 TRANSMISSIONS

A *transmission* is a gear train designed to operate at several different output speeds. This is accomplished by changing the gears that are in mesh in the gear train. Since the automotive three-speed manual transmission is a common example of a transmission, we shall explain its functioning to illustrate how different output speeds are obtained.

Figure 15–11 is a schematic drawing of a three-speed automotive manual transmission. Input power from the engine crankshaft is taken at gear A through a clutch controlled by the operator of the car (this clutch is not shown). The gearshift lever (not shown) moves the clutch and the gear F as indicated in Figure 15–11 to obtain the various output speeds. It is important to note the following gear-to-shaft relationships:

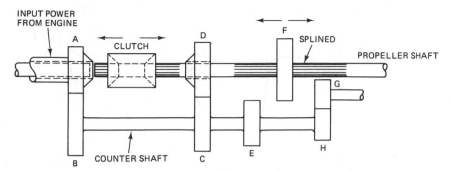

FIGURE 15–11.   Three-speed automatic transmission.

Action	Power flow	Ratio
F shifts left, engaging E	Low gear: $A \to B \to E \to F$	$\left(\dfrac{n_A}{n_B}\right)\left(\dfrac{n_E}{n_F}\right)$
Clutch shifts right, fixing D on propeller shaft	Second gear: $A \to B \to C \to D$	$\left(\dfrac{n_A}{n_B}\right)\left(\dfrac{n_C}{n_D}\right)$
Clutch shifts left, fixing A and propeller shaft	Third gear: $A \to$ propeller shaft	1 : 1
F shifts right, engaging G	Reverse: $A \to B \to H \to G \to F$	$\left(\dfrac{n_A}{n_B}\right)\left(\dfrac{n_H}{n_F}\right)$

1. Gears A and D are free to rotate on the shaft until restrained by the clutch.
2. The clutch and gear F are splined to the propeller shaft but can be moved axially.
3. Gears B, C, E, and H are fixed to the countershaft.

The gear ratios for each output speed are shown in Figure 15–11. They are obtained by applying Equation 15–5 to the gear train. The 1 to 1 ratio of the third gear results when a direct connection between gear A and the propeller shaft is made by the clutch.

► PROBLEMS

15–1   Classify by type all the gear trains in Figure 15–12.

15–2   Locate and identify all the idler gears in Figure 15–12.

15–3   Determine the direction of rotation of gear B in Figure 15–12.

15–4   Using a simple ordinary gear train with five gears, prove that Equation 15–3 holds for this condition.

15–5   Calculate the velocity ratio for the gear train in Figure 15–12(a).

15–6   Calculate the velocity ratio for the gear train in Figure 15–12(b).

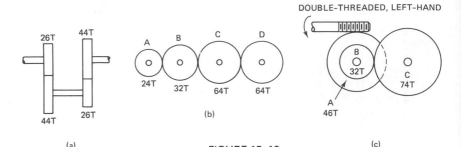

FIGURE 15-12

15-7 Calculate the velocity ratio for the gear train of Figure 15–12(c). If the worm rotates at 740 rpm, what is the output speed?

15-8 In Figure 15–13, what is the speed and direction of rotation of gear *D*?

15-9 A simple ordinary gear train has 33 teeth on the input gear and 75 on the last driven gear. What is the velocity ratio?

15-10 Find the speed and direction of rotation of gear *D* in Figure 15–14.

15-11 A three-speed automotive transmission has gears with the following number of teeth. The gear position is that shown in Figure 15–11.

Gear *A*: 20 T   Gear *B*: 36 T   Gear *C*: 25 T   Gear *D*: 21 T
Gear *E*: 15 T   Gear *F*: 27 T   Gear *H*: 14 T

Calculate the velocity ratio in each gear, low, second, high, and reverse.

15-12 How many turns of the worm in Figure 15–15 are required to raise the weight 1 ft? What direction must the worm be turned?

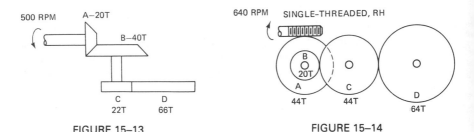

FIGURE 15-13   FIGURE 15-14

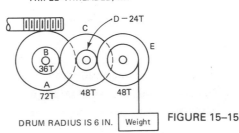

238   DRUM RADIUS IS 6 IN. [Weight]   FIGURE 15-15

CHAPTER 16
PLANETARY GEAR TRAINS

16–1   INTRODUCTION

In the *planetary gear train*, one or more gear axes move relative to the frame. Planetary gearing is also called *epicyclic gearing*; its main advantage over ordinary gearing is the larger speed reduction that can be obtained in a given amount of space as compared with ordinary gear trains. Figure 16–1 shows a typical commercial planetary gear speed reducer.

   Whereas the automotive manual transmission uses the ordinary gear train, most automatic transmissions using hydraulic torque converters use planetary gearing to achieve the low, second, drive, and reverse gears available with the transmission. Figure

FIGURE 16–1. Planetary speed reducer. (Courtesy Winsmith Division of UMC Industries, Inc.)

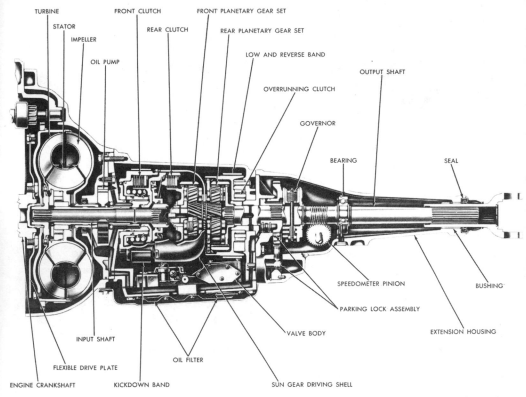

FIGURE 16–2. TorqueFlite Transmission and Torque Converter (A-904). (Courtesy Chrysler Corporation)

16–2 shows the cross section of an automatic transmission with its associated planetary gearing.

In the planetary gear train, gears having axes not fixed to the frame rotate about a gear called a *sun gear*. The gears rotating about this sun gear are called *planet gears*, thus giving rise to the term planetary gear train. This relationship is shown in Figure 16–3; the large gear is the sun gear. Its axis is fixed, and the gear itself is also fixed so that it does not rotate. The planet gear is held in mesh with the sun by the *carrier* (also called an *arm*). In this case the carrier rotates about the fixed axis of the sun gear, forcing the planet gear to roll on the surface of the sun gear.

16–2 RELATIVE AND ABSOLUTE MOTION

Consider the planetary gear train of Figure 16–4. The train is similar to that of Figure 16–3 except that pitch diameters, and correspondingly the circumferences, are in the ratio of 1 to 2. The carrier

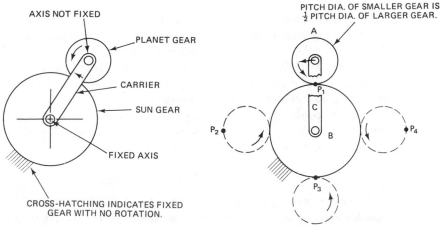

FIGURE 16–3. Planet and sun gear.

FIGURE 16–4. Travel of point *P* relative to carrier.

rotates counterclockwise, forcing gear *A* to roll without slip on gear *B*.

If a point *P* is selected on the pitch circle of gear *A* with initial position of P_1 as shown. The circumference of *A* is one half the circumference of *B*, and *A* will have completed one revolution as it rolls from P_1 to the P_3 position. Two complete revolutions of *A* are completed when *A* is back in its original position. This can be proved experimentally by using two circular disks with diameters in the ratio of 1 to 2, marking a point on the smaller, and rolling it around the larger.

The motion of point *P* described in the preceding paragraph is its motion *relative* to the carrier. Additional motion exists, as indicated by Figure 16–5. Here gear *A* is assumed to be

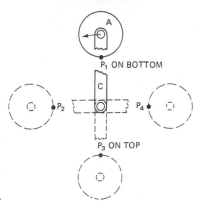

FIGURE 16–5. Gear *A* fixed to carrier.

fixed to the carrier as it rotates through one revolution. The same point P is shown at positions 1, 2, 3, and 4. Note that, at position 1, P_1 is on the *bottom* of the gear, whereas at position 2, P_2 is on the *top* of the gear. For this to occur, gear A must have rotated one half of a revolution. Therefore, when gear A returns to its original position, one complete revolution of A has taken place.

By applying the principles of Chapter 5 relating to absolute and relative velocities, the following equation can be developed for *angular velocity* of bodies in plane motion. It is

$$\omega_1 = \omega_2 + \omega_{1/2} \qquad (16\text{–}1)$$

The angular velocity of a body having plane motion is equal to the angular velocity of a second body plus the angular velocity of the first body *relative* to the second.

Applying this to Figures 16–4 and 16–5 and using subscript C for the carrier, we have

$$\omega_A = \omega_C + \omega_{A/C} \qquad (16\text{–}2)$$

In Equation 16–2, sign conventions for direction of rotation must be used. We consider counterclockwise to be plus (+) and clockwise rotation to be minus (−).

Substituting the values found for Figures 16–4 and 16–5, we have

$$\omega_A = +(1) + (+2)$$

$$= +3$$

The conversion from revolutions to revolutions per minute is made by assuming that the rotation occurs in a time interval of 1 min, thus allowing its use as an angular velocity term.

The gear train of Figure 16–4 can be converted to an ordinary gear train by releasing gear B and fixing the carrier, C. Since the ratio of diameters is 1 to 2, gear A turns 2 rpm when gear B turns 1 rpm. The velocity ratio is $\frac{1}{2}$. For the planetary train, gear A turns 3 rpm and the velocity ratio is $\frac{1}{3}$. *This illustrates the greater speed reduction available from planetary gearing as compared to ordinary gear trains.*

Equations 16–1 and 16–2 can be rewritten in a more generalized form for greater usefulness. Using the subscripts G to represent any planet gear and C to represent the carrier,

$$\omega_G = \omega_C + \omega_{G/C} \qquad (16\text{–}3)$$

16–3 ANALYSIS OF PLANETARY GEAR TRAINS

The motions of individual gears in a planetary gear train cannot be readily visualized, nor can ratios be determined with the same ease as with ordinary gear trains. However, methods are available that allow accurate determination of both of these. The principles of relative and absolute motion discussed in Section 16–2 are used in these methods.

Figures 16–3 and 16–4 show planetary gear sets with one fixed gear. Planetary gear sets also exist with no fixed gear, and the motion relationship is determined by the input and output of the gear set. The methods for analyzing these two vary somewhat. Since having one fixed gear usually makes motion analysis simpler, we start with the fixed gear train.

Figure 16–6 shows a planetary gear train with one fixed gear. The train is similar to that of Figure 16–4 except for the additional gear, *D*. Gear *D* is an internal gear meshing with gear *A*. The carrier *C* rotates as shown. We are to find the number of revolutions of each gear and the direction of its rotation resulting from one revolution of the carrier.

The procedure consists of three steps. The first step is the consideration of the train with all members locked together and rotated as a unit. This was used in Figure 16–5. The second step is to convert the train to an ordinary train by locking the carrier and releasing the fixed gear. In both of these steps, the number of revolutions of each gear is determined using gear tooth ratios. The final step is the application of Equation 16–3. These three steps are illustrated in Figure 16–7, using the gear train of Figure 16–6. The procedure is also described in the following.

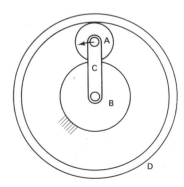

FIGURE 16–6

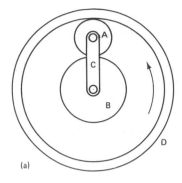 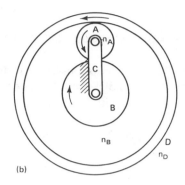

(a) (b)

(a) Step 1: Locked—All gears rotate + 1 revolution

(b) Step 2: Ordinary gear train—Lock the carrier and release the fixed
gear; rotate the fixed gear − 1 revolution; calculate the turns of the
other gears

(c) Step 3: Add algebraically the number of turns of each member

FIGURE 16–7.   Planetary gear train with fixed gear.

Step 1

Lock all gears and the carrier together and rotate the unit+1 revolution. All members then rotate +1 revolution.

Step 2

Release the fixed gear and lock the carrier. Rotate the fixed gear −1 revolution and calculate the number of turns that each other gear makes, also noting the rotation direction and indicating by appropriate sign.

Step 3

Algebraically add the results from steps 1 and 2 for each member. The direction of rotation of each gear is indicated by its sign.

The procedure is best applied using a tabular form to list each member of the train and the number of revolutions made by each member. We illustrate it in the following examples.

EXAMPLE

The gear train in Figure 16–6 has the following number of teeth on each gear.

$$n_A = \ 26$$

$$n_B = 40$$

$$n_D = 120$$

Find the number of turns that each gear makes for one revolution of the carrier.

Solution

The solution uses a table set up as follows. Each step is listed, as is each train member.

Number of Turns

Step	C	B	A	D
1	+1	+1	+1	+1
2	0	−1	$+(1)(\frac{40}{26})$	$+(\frac{40}{26})(\frac{26}{120})$
3	+1	0	$+2\frac{14}{26}$	$+1\frac{40}{120}$

The results can be interpreted as follows:

1. The carrier, C, has turned one revolution.
2. Gear B is stationary, indicated by the 0 turns in the answer.
3. Gear A has turned $2\frac{14}{26}$, or $2\frac{7}{13}$, revolutions for the one revolution of the carrier. The direction of rotation is the same as the carrier.
4. Gear D has turned $1\frac{40}{120}$, or $1\frac{1}{3}$, revolutions for one revolution of the carrier. The direction of rotation is the same as the carrier.

In step 2 the direction of rotation is determined in the same manner as used for the ordinary gear train. The arrows indicating rotation are shown in Figure 16–7(b).

Calculation of the gear revolutions in step 2 is done by starting with the first known gear and then proceeding with the next gear in mesh with it, using the basic equation for velocity ratio.

The equations are

For A,

$$\frac{\omega_A}{\omega_B} = \frac{n_B}{n_A}$$

$$\omega_A = \omega_B\left(\frac{n_B}{n_A}\right)$$

Substituting,

$$\omega_A = 1\left(\tfrac{40}{26}\right)$$

for D,

$$\frac{\omega_D}{\omega_A} = \frac{n_A}{n_D}$$

$$\omega_D = \omega_A\left(\frac{n_A}{n_D}\right)$$

Substituting,

$$\omega_D = 1\left(\tfrac{40}{26}\right)\left(\tfrac{26}{120}\right)$$

$$= \left(\tfrac{40}{26}\right)\left(\tfrac{26}{120}\right)$$

The answers have not been reduced to lowest terms but are left in the form used in the table.

EXAMPLE

The gear train in Figure 16–8 is the same as the train shown in Figure 16–6 except that gear D is the fixed gear. Gear B is the driver and rotates at 1000 rpm clockwise. Carrier C is the output. Find its angular velocity and direction of rotation.

Solution

The table listing the number of turns of each member is set up as before.

Number of Turns

Step	C	D	A	B
1	+1	+1	+1	+1
2	0	−1	$-(1)\left(\tfrac{120}{26}\right)$	$+\left(\tfrac{120}{26}\right)\left(\tfrac{26}{40}\right)$
3	+1	0	$-3\tfrac{16}{26}$	+4

The velocity ratio is the speed of the driven member divided by the speed of the driver. Gear B is the driver, and the carrier C is the driven member. The velocity ratio is then $\tfrac{1}{4}$, and the output velocity is $\left(\tfrac{1}{4}\right)(1000)$, or 250 rpm. The signs indicate that the direction of rotation is the same for both members, and therefore the carrier C rotates clockwise.

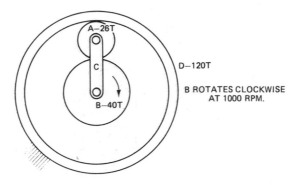

FIGURE 16–8

Note that, in setting up the table, the fixed gear D followed the carrier C in order. The order of the two remaining gears then is obtained by proceeding from D to A and from A to B.

When a planetary train has no fixed gear, the method for analysis has to be modified somewhat. In general, with no fixed gear, it is necessary to know the speeds of two members of the train. The same requirement actually exists for a train with a fixed gear, since fixing the gear gives one velocity and the other was assumed by using one revolution in the solution. Equation 16–3, relating to the relative velocity of gear to carrier, is also used in the solution.

Consider the gear train shown in Figure 16–9. The input shaft rotates at 100 rpm clockwise as viewed from the left. Gears A and D are fixed to this shaft and thus rotate at the same speed. A and D drive B and E of the planetary train. The carrier C and gear B are fixed to the intermediate shaft, but gear E is free to rotate about the intermediate shaft. Gear F is a planet that drives the internal gear G. It is desired to find the output speed.

The modified procedure is described in the following:

Step 1

In each column of the table, enter the angular velocity of the carrier using the correct sign.

Step 2

Using Equation 16–3, calculate the velocity relative to the carrier of any gear whose absolute angular velocity is known. Equation 16–3

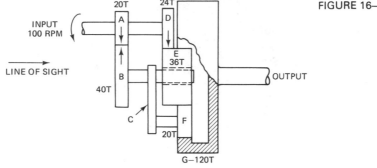

FIGURE 16–9

$$\omega_{G/C} = \omega_G - \omega_C$$

Then fix the carrier and rotate the gear selected at the velocity calculated from Equation 16–3, being sure that the correct sign is used. Calculate the velocity of the other gears, using the methods applicable to ordinary gear trains.

Step 3

Add the results of steps 1 and 2 algebraically.

The application of these steps to the problem in Figure 16–9 is developed in the following solution.

Solution

Gears A and D provide the two required speeds for the planetary gearing but are not part of the planetary set. Since the carrier and B are on the same shaft, the angular velocity of the carrier is the same as that of B. The velocity of B is

$$\omega_B = -\tfrac{20}{40}(100) = +50 \text{ rpm}$$

Then

$$\omega_C = +50 \text{ rpm}$$

The absolute angular velocity of E is

$$\omega_E = +\tfrac{24}{36}(100) = +66\tfrac{2}{3} \text{ rpm}$$

Equation 16–3 is now applied for gear E.

$$\omega_{E/C} = \omega_E - \omega_C$$

Substituting, we have

$$\omega_{E/C} = (+66\tfrac{2}{3}) - (+50) = +16\tfrac{2}{3} \text{ rpm}$$

We can now set up the table.

Number of Turns

Step	C	E	F	G
1	+50	+50	+50	+50
2	0	$+16\frac{2}{3}$	$-(16\frac{2}{3})(\frac{36}{20})$	$-(30)(\frac{20}{120})$
3	+50	$+66\frac{2}{3}$	+20	+45

In step 2 the velocity of E when reduced is 30, and this number is used in the term for G in step 2.

The speed of gear G, the output, is 45 rpm in a counterclockwise direction.

16–4 GEAR DIFFERENTIAL

The *gear differential* is a planetary gear train design that has the capability of adding or subtracting angular displacements. It uses bevel gears and has no fixed gear. Figure 16–10 shows a differential. The sun gears, A and B, are of the same size, while the planet P may be of a different size.

The relationships between the angular velocities of the sun gears and the carrier may be found by using the procedure for planetary gear trains with no fixed gear. Letting ω_A, ω_B, and ω_C be the angular velocities of gears A and B and the carrier, the table is set up as follows.

FIGURE 16–10.   Gear differential.

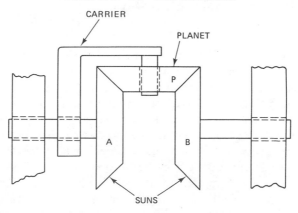

Comparison of Angular Velocities

Step	C	A	B
1	ω_C	ω_C	ω_C
2	0	$+(\omega_A - \omega_C)$	$-(\omega_A - \omega_C)$
3	ω_C	ω_A	$2\omega_C - \omega_A$

In step 1 the angular velocity of the carrier, ω_C, is entered for all three members. Step 2 requires fixing the carrier and considering the train as an ordinary gear train. Equation 16–3 is then applied to find the velocity of A relative to the carrier. It is

$$\omega_{A/C} = \omega_A - \omega_C$$

The term $(\omega_A - \omega_C)$ is now listed under A for step 2. Note that the sign is plus, the same as that assumed for the carrier rotation.

In step 2 the train is an ordinary train with A now driving the planet and gear B. Since B is the same size as A, its angular velocity relative to the carrier is the same as the angular velocity of A. However, its direction of rotation is opposite to that of A and requires a minus sign. Mathematically,

$$\omega_{A/C} = -\omega_{B/C}$$

This relationship allows the substitution of the term $-(\omega_A - \omega_B)$ under B in step 2.

When all steps are added, the results are as shown. If we take the term for the angular velocity of B and equate it to ω_B, we have

$$\omega_B = 2\omega_C - \omega_A \tag{16-4}$$

Rearranged, Equation 16–4 is

$$\omega_A + \omega_B = 2\omega_C \tag{16-5}$$

Equation 16–5 states that the sum of the two angular velocities of the sun gear is always twice the angular velocity of the carrier. If the units of time are removed from Equation 16–5 by dividing through both sides of the equation, angular displacement units remain. The equation therefore applies for displacement, allowing the differential to be used for adding and subtracting displacements. Because of this property, differentials have been used in mechanical computers for solving equations in the form of Equation 16–5.

A common example of the application of the differential is the automobile rear axle shown in Figure 16–11. The bevel gear differential here is similar to that shown in Figure 16–10 except for the use of two planets instead of one. The two planets are required to provide a symmetrical assembly that balances the loading on the gears; the additional planet does not change the assembly kinematically.

The pinion drives the hypoid ring, which also functions as the carrier. The two outputs are the right and left axles and Equation 16–5 applies, making the sum of the two angular velocities equal to twice the ring gear velocity. As one axle slows down, as when turning a corner, the other axle speeds up by the same amount to make the sum always equal to twice the ring gear speed.

▶ PROBLEMS

16–1 What is the main advantage of planetary gearing over the ordinary gear train?

16–2 In Figure 16–12, gear *A* is fixed. Find the ratio of the angular velocity of the carrier to gear *B*. Does gear *B* rotate in the same or opposite direction as the carrier?

16–3 If the carrier in Figure 16–12 rotates clockwise at 600 rpm, find the angular velocity of gear *B* and its direction of rotation.

16–4 If gear *B* is made the fixed gear instead of gear *A* in Figure 16–12, what is the ratio of the carrier velocity to gear *A*? What is the effect of changing the fixed gear from *A* to *B*?

16–5 The carrier of the gear train in Figure 16–13 rotates in the direction shown at 750 rpm. What is the speed of gear *A* and its direction of rotation?

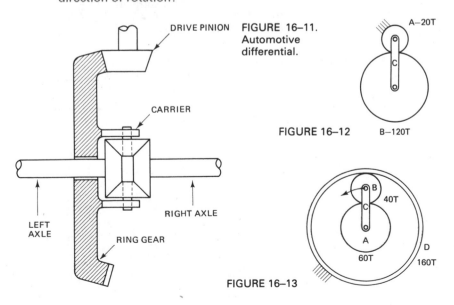

DRIVE PINION

FIGURE 16–11.
Automotive
differential.

CARRIER

RIGHT AXLE

LEFT
AXLE

RING GEAR

A–20T

C

FIGURE 16–12   B–120T

B
40T
C
A
60T   D
            160T

FIGURE 16–13

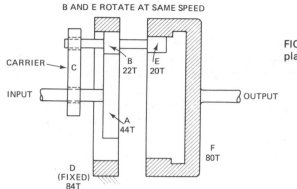

B AND E ROTATE AT SAME SPEED

CARRIER

INPUT

C

B
22T

E
20T

A
44T

D
(FIXED)
84T

F
80T

OUTPUT

FIGURE 16–14. Compound planetary gear train.

16–6 A *compound* planetary gear train is sometimes used in speed reducers where a very high speed reduction is required. Figure 16–14 shows such a gear train. The fixed gear is gear *D*, one of the internal gears. The other internal gear rotates and is the output. Gears *B* and *E* are fixed to the same shaft and rotate at the same speed. If the input speed is 100 rpm, find the output speed.

16–7 To find the angular velocities of the gears of a planetary gear set with no fixed gears, the angular velocities of _____ members of the train must be known. (Fill in the correct answer.)

16–8 In the gear differential shown in Figure 16–10, the carrier rotates at 500 rpm counterclockwise when viewed from the left. Sun gears *A* and *B* also rotate counterclockwise, and both sun gears have the same speed. What is the speed of the planet *P*?

16–9 The carrier *C* of the gear train shown in Figure 16–15 rotates at 60 rpm ccw. Gear *A* rotates at 10 rpm cw. Find the speeds of gears *B* and *D*.

16–10 The right axle of the automotive differential in Figure 16–11 rotates at +65 rpm and the left axle at +75 rpm. What is the speed of the ring gear?

16–11 If the rear axle ratio of the differential in Problem 16–10 is 3.3 to 1, find the drive pinion speed. Use the data from Problem 16–10 for your solution.

16–12 In Figure 16–10, gear *A* rotates at +10 rpm and *B* rotates at −10 rpm. What is the speed of the carrier?

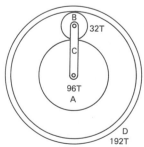

B
32T

C

96T
A

D
192T

FIGURE 16–15

CHAPTER 17
BELT, CHAIN, AND OTHER DRIVES

17-1 INTRODUCTION

Belt and chain drives kinematically are similar to gear drives, with velocity ratios inversely proportional to pitch diameters. Because of their widespread use, we shall cover belt and chain drives in this chapter, in addition to discussing some of the other types of drives that are not as well known.

17-2 BELT DRIVES

The V-belt drive shown in Figure 17-1 is typical of a large number of belt drives. The load-carrying capacity of the drive is increased by using three belts instead of one. Increasing the load capacity of belt

FIGURE 17-1. V-belt drive. (Courtesy Browning Manufacturing Division, Emerson Electric Co.)

253

NEUTRAL SURFACE OF BELT.

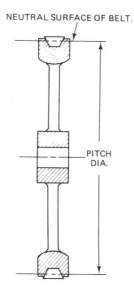

PITCH
DIA.

FIGURE 17–3. Belt and pulley velocity.

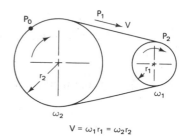

$$V = \omega_1 r_1 = \omega_2 r_2$$

FIGURE 17–2. Pitch diameter of belt and pulley.

drive allows the belt drive to approach the load-carrying ability of gearing, and thus removes one of the disadvantages of the belt drive as compared to gears. Belt drives have the advantage generally of lower cost than gearing, have some shock-absorbing ability, which gears lack, and allow the use of connecting shafts that are too far apart for two gears to mate.

The velocity ratio of belt drives is determined by applying the equation $v = \omega r$ (Equation 4–15) in a manner similar to that for gears. Consider the section of the pulley and belt shown in Figure 17–2. The *pitch diameter* of the pulley (for the V-belt, the pulley is commonly called a *sheave*) is a theoretical diameter between the belt surfaces that are not stressed by the bending of the belt. It is called the *neutral surface* and is identified in Figure 17–2. The location of this surface depends on the belt and sheave design and cannot be readily calculated. Instead, manufacturers catalog data must be used to find the pitch diameter of a sheave.

The V-belt drive depends on frictional forces exerted by wedging action of the belt in order to transmit the load. Some slip occurs despite the wedging, however, and calculations for velocity ratios are not as precise as those for gears. Figure 17–3 may be used to illustrate how Equation 4–15 is applied to find velocity ratios. The belt connects two pulleys, one of which is the driver. If it is assumed that the belt is inelastic and that no slip occurs, then the linear velocity has the same magnitude at all points. Thus, at points P_0, P_1, and P_2 the magnitude of the velocity is the same. Now P_0 and

P_2 are located on the pitch circles of the two pulleys, and Equation 4–15 can be used to write the following:

$$v = \omega_1 r_1 = \omega_2 r_2$$

Rearranging and substituting $D/2$ for r gives

$$\frac{\omega_2}{\omega_1} = \frac{D_1}{D_2} \qquad (17\text{–}1)$$

Equation 17–1 is the same as Equation 11–10, developed for gears, and is the *angular velocity ratio* of the two pulleys. Because of the existence of some elasticity in the belt and some slip, the equation should be considered approximate.

In years past most factories depended on *flat belts*, usually leather, to transmit power from overhead line shafts to shafting that drove the machines. Flat belting of this type has been replaced by individual drives, but for specialized applications flat belts continue to be used. Light fabric belts are used for high-speed drives where loads are light and also where transmission of vibration is not desired. The pitch diameter of a flat-belt pulley is equal to the pulley outside diameter plus one belt thickness.

Toothed pulleys and *toothed belts* are used in applications where it is desired to eliminate slip. Figure 17–4 shows typical design details of the toothed pulley and belt. Toothed belts and pulleys are accurate enough to allow their use as timing drives. *Double-toothed belts* are also manufactured. These are flat belts with teeth on both sides of the belt. Double-toothed belts allow the use of drives such as those shown in Figure 17–5.

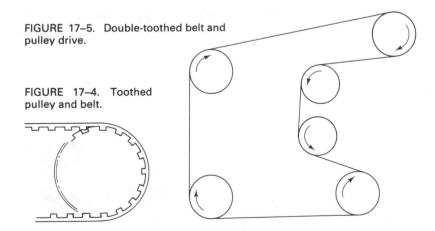

FIGURE 17–5. Double-toothed belt and pulley drive.

FIGURE 17–4. Toothed pulley and belt.

17–3 CHAIN DRIVES

Chain drives provide high load capacity and allow the spacing of shafts a distance apart. Toothed sprockets, as shown in Figure 17–6, prevent slip in the drive. Equation 17–1 applies for the velocity ratio. In addition, the velocity ratio is inversely proportional to the number of teeth on the sprockets. Thus,

$$\frac{\omega_2}{\omega_1} = \frac{n_1}{n_2}$$ (17–2)

Various manufacturers offer different chain designs and different degrees of precision. *Roller chain* is probably the most used type. The name is derived from the fact that the chain element in contact with the sprocket tooth is a roller.

Silent chain drives use an inverted tooth design, such as shown in Figure 17–7. The chain tooth maintains contact with the sprocket without sliding; as a result, the chain is quieter than the roller chain.

17–4 VARIABLE-SPEED DRIVES

A common method for changing the velocity ratio for V-belt drives is to use stepped pulleys as shown in Figure 17–8. This method requires stopping the machine and transferring the belt to the desired set of pulleys. Usually one shaft can be temporarily shifted out of position in order for the belt to be moved to the other set of pulleys.

For many machines the stepped pulley is adequate for speed changing. For some applications, however, it is necessary to

FIGURE 17–6. Sprockets for chain drive. (Courtesy Browning Manufacturing Division, Emerson Electric Co.)

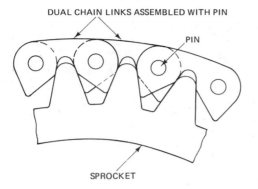

DUAL CHAIN LINKS ASSEMBLED WITH PIN

PIN

SPROCKET

FIGURE 17–7. Silent chain.

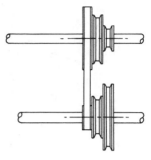

FIGURE 17–8. Stepped pulleys for changing speeds.

change the speed of an output while the mechanism drive is in motion. *Variable speed drives* have been developed to accomplish this. One of these is shown in Figure 17–9.

The principle behind the design of most variable-speed drives lies in the application of Equation 4–15, $v = \omega r$. A change in the radius of a rotating member changes the velocity of another member in contact with it. Most variable drives therefore are designed so that the radius may be changed while the drive is rotating.

Frictional forces exerted through the balls of the variable-speed drive in Figure 17–9 drive the output shaft. The change in the radius that provides the speed change is accomplished by changing the angle of the axis of rotation of the ball. Figure 17–10 shows how this works. Two disks of the same diameter contact the ball, which is driven by the input disk. At the point of contact the effective radius

FIGURE 17–9. Winkopp variable-speed drive. (Courtesy Winsmith Division of UMC Industries, Inc.)

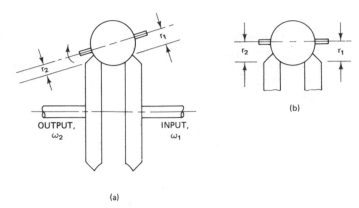

FIGURE 17–10. Variable-speed ball-disk drive.

measured from the ball axis is r_1 on the input side and r_2 on the output side. The velocity ratio is

$$\frac{\omega_2}{\omega_1} = \frac{r_2}{r_1} \qquad (17\text{--}3)$$

Note that the ratio is directly proportional to the radii, not inversely proportional.

In Figure 17–10(a), the ball axis is shown at an angle giving unequal values of r_1 and r_2. If the ball axis is parallel to the input shaft axis, as shown in Figure 17–10(b), then r_1 equals r_2. The ball axis is moved by cams, which are adjusted by a worm and gear; in Figure 17–9 the gear may be seen. The frictional forces that can be obtained with the ball and disk are lower than those available with some other drives, particularly the V-belt. The load capacity is therefore limited.

The V-belt variable-speed drive is available where higher loads require more capacity. Speed change is accomplished by changing the pitch radius simultaneously on the driver and driven sheaves. Its principle is illustrated in Figure 17–11. Each sheave is constructed in halves so that the two halves can be moved axially toward and away from each other. The pitch radius changes accordingly, as indicated by the enlarged section of sheave and belt in Figure 17–11(b). The velocity ratio of the variable-speed V-belt drive is calculated in the same manner as for the straight V-belt drive, with the radii values being those that exist at a particular adjustment.

17–5 MISCELLANEOUS DRIVES

There are other drive types too numerous to discuss in this text. Some of these are *wheel and disc drives, planetary cone transmission, flexible spline drives,* and *screw actuators.* The reader is referred to more advanced texts for information on these.

► PROBLEMS

17–1 Give two advantages of V-belt drives over gearing.

17–2 A V-belt drive is to be designed using a drive pulley with a 6.75-in. pitch diameter. The driven pulley has a pitch diameter of 13.86 in. What is the velocity ratio?

17–3 A flat-belt drive uses a belt $\frac{1}{16}$ in. thick. The drive pulley outside diameter is 2 in., and the driven pulley is 3 in. in diameter. Find the velocity ratio.

17–4 A chain drive uses a sprocket with 18 teeth for the driver. The driven sprocket has 30 teeth. What is the velocity ratio?

17–5 The silent chain design uses _____. (Select the correct answer.) (a) rollers for sprocket teeth contact (b) a variable pitch sprocket (c) an inverted tooth design for quietness (d) none of these

FIGURE 17–11. Variable-speed V-belt drive.

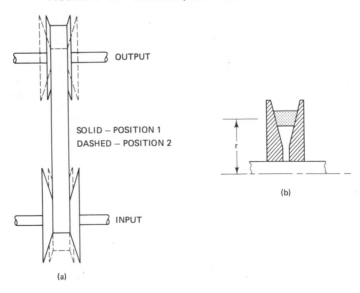

OUTPUT

SOLID – POSITION 1
DASHED – POSITION 2

r

(b)

INPUT

(a)

17–6 Equation 17–3 states that the velocity ratio of the variable-speed ball-disk drive is proportional to the effective radii of the ball at the contact points. Prove that this is true.

17–7 Many variable-speed drives accomplish a speed change by changing the _____ of a rotating member.

17–8 A variable-speed V-belt drive uses a driven sheave with an 8-in. pitch diameter and a driver with pitch diameter adjustable from 4.9 to 9.7 in. The center distance can be adjusted to accommodate the change diameter of the driver. What is the minimum and maximum velocity ratio?

APPENDIX
ANSWERS TO EVEN-NUMBERED PROBLEMS

CHAPTER 1

1–2 The Wankel rotary engine uses a 3-lobe rotary shaft, which compresses the gas mixture.

1–4 A mechanism is a combination of machine elements designed and assembled to produce certain specific motions.

1–6 One, because of wear considerations, and two, for acceleration and force determination.

1–8 So that forces, and consequently stresses, in the links may be determined.

CHAPTER 2

2–2 The wheel of an automobile as the automobile moves on a highway.

2–4 Combined translation and rotation.

2–6 38.8 in. up and to the left at an angle of 11.4° with the horizontal.

2–8 40.5 ft down and to the right at an angle of $13\frac{1}{4}°$ with the horizontal.

2–10 North: 28.5 miles
West: 56.5 miles

2–12 C = 15 cm
D = 33 cm

2–14 E and F
A and B

CHAPTER 3

3–2 (a) No (c) No
(b) Yes (d) No

3–6 Not feasible

3–8 Incorrect

3–10   (a) Double rocker   (c) Indefinite motion
      (b) Drag link   (d) Crank rocker

CHAPTER 4

4–2   Yes   4–4   2.38 miles

4–6   92.9 ft/sec   4–8   9.6 rad/sec^2

4–10   5.8 min   4–12   255 rpm

4–14   12 rpm   4–16   840 rpm

4–18   1273 rpm

CHAPTER 5

5–2   (a) $v_A = 5.54$ mph 22.5° up and to the right
         $v_B = 4.24$ mph 45° down and to the right
         $v_C = 4.24$ mph 45° up and to the right
    (b) $v_{A/B} = 8.1$ ft/sec $67\frac{1}{2}$° up and to the right
         $v_{A/C} = 3.4$ ft/sec $22\frac{3}{4}$° down and to the right

5–4   319 cm/sec perpendicular to the link

5–6 $v_{PIN/SLOT} = 60.6$ in/sec to the left
      $v_{SLIDER} = 121.2$ in/sec 60° down and to the left

5–8 $v_A = 209$ in/sec 60° down and to the left
      $v_B = 76$ in/sec to the left

5–10 $v_A = 42.8$ ft/sec 30° up and to the left
       $v_B = 41$ ft/sec 31° up and to the left
       $v_C = 40$ ft/sec $26\frac{1}{2}$° up and to the left
       $v_D = 43\frac{1}{2}$ ft/sec to the left

5–16   701 cm/sec   5–20   1 rad/sec

CHAPTER 6

6–2 $a_A^t = 23.3$ ft/sec^2, $a_A = 12794$ ft/sec^2
     $a_B^t = 30$ ft/sec^2, $a_B = 16449$ ft/sec^2

6–4   12324 ft/sec^2 6–6   82.2 ft/sec^2

6–8   98596 cm/sec^2 6–10 $a_A = 363.3$ ft/sec^2 down
                               $a_B = 0$

6–12   5.8 ft/sec^2 to left   6–14 $a_B = 8800$ cm/sec^2
                                  $a_C = 3500$ cm/sec^2

6–16 $a_C = 0$ 6–18   39270 ft/min^2

6–20   658 in/sec^2

CHAPTER 7

7–2　(a) AC = 5.24 in.　　　　7–4　7270 ft/sec^2
　　　(b) 2.1

7–6　72.0 ft/sec　　　　　　7–8　14.7 in.

7–10　13.2 in.　　　　　　　7–12　12 in.

7–14　Infinite: O_2B would be parallel to O_1O_2 in this case; this linkage
　　　would not work.

7–16　1.27　　　　　　　　　7–18　22.5 m/sec^2

CHAPTER 8

8–2　The wheel indexes each time the arm is actuated, instead of
　　　once for each two times the arm is actuated.

8–4　0°　　　　　　　　　8–6　25%

8–8　Mainspring

8–10　The drive is by pins, as in the Geneva mechanism.

CHAPTER 9

9–2　Plate cam

9–4　Roller cam

9–6　Constant velocity, constant acceleration, cycloidal motion,
　　　and simple harmonic motion are used to describe cams.

9–8　Except for special types such as the bar cam, the profile must
　　　be circular.

9–10　30°

9–12　Cam surface

CHAPTER 10

10–16　Modified constant velocity

CHAPTER 11

11–2　Bevel

11–4　The involute is the path traced by a point on a string as the
　　　string is unwound from a circle.

11–6 Diametral pitch

11–10 795 rpm

11–12 6 diametral pitch,
36 teeth

11–16 32 and 48

11–8 8 in. pitch dia.,
0.785 circular pitch

11–14 (a) 7 in. and 24 in.
(b) 96
(c) $15\frac{1}{2}$ in.

CHAPTER 12

12–2 $14\frac{1}{2}°$

12–4 2.125 in., catalog 2.12 in.

12–6 Interference occurs with the reduced pinion diameter and the effect is to increase the probability of interference.

12–8 Addendum: 0.0833 (same as the $14\frac{1}{2}°$ gear)
Dedendum: 0.1042 (greater than the $14\frac{1}{2}°$)
Whole depth: 0.1875 (greater than the $14\frac{1}{2}°$)
Clearance: 0.0208 (greater than the $14\frac{1}{2}°$)

12–10 An 8-pitch, 32-tooth gear should be used.

12–12 Addendum: 0.125
Dedendum: 0.156
Clearance: 0.031
Whole depth: 0.281

CHAPTER 13

13–2 The thrust force which is exerted axially on the shaft.

13–4 Left hand

13–6 (a) 0.393
(b) 0.322
(c) 9.756

13–8 Two satisfactory gears are: diametral pitch—10, 42 and 168 teeth, helix angle—30°. Pinion is left hand, gear is right hand. Note that other combinations will also work.

13–10 90°

13–12 One combination is 10 diametral pitch, 25 and 30 teeth. Other combinations exist.

CHAPTER 14

14–2 The intersection of the back cone with the pitch cone.

14–4 It allowed the drive shaft to be lowered and at the same time was quieter and carried more load than the straight bevel gear.

14–6   4.125 in. and 5.5 in.   *14–8*   72.48°

14–10 62.5 rpm

CHAPTER 15

15–2   Gear C in Fig. 15–12 (b)   *15–6*   0.375

15–8   83.35 rpm cw   *15–10* 4.55 rpm cw

15–12 20.4 turns

CHAPTER 16

16–2   1 to 1$\frac{1}{16}$, same direction   *16–4*   1 to 7; a higher velocity is obtained.

16–6   25.78 rpm   *16–8*   0 rpm

16–10 70 rpm   *16–12* 0 rpm

CHAPTER 17

17–2   0.487   *17–4*   0.6

17–8   0.6125 and 1.2125

INDEX